AF572579

Purification Tools for Monoclonal Antibodies

by Pete Gagnon

Validated Biosystems

Purification Tools for Monoclonal Antibodies
by Pete Gagnon

First published in 1996 by Validated Biosystems, Inc.,
5800 N. Kolb Road, Suite 5127, Tucson, AZ 85750

Library of Congress Catalog Card Number
96-90448

International Standard Book Number
0-9653515-9-9

Foreword

"All truths are easy to understand once they are discovered; the point is to discover them"
—Galileo

Out of all the biotechnology products in development, in clinical trials, and on the market, monoclonal antibodies are the most numerous. They continue to be at the forefront of every new field of endeavor, including human and veterinary healthcare, agriculture, forensics, and environmental monitoring. Even as competing product classes arise, technology improvements and expanding applications guarantee that monoclonals will remain a vital force in the continuing growth of the industry.

This places enormous pressure on purification process designers. In today's cost conscious environment, economical purification schemes that support the requisite product quality and meet regulatory expectations are critical for success. Agressive time-to-market calendars demand that they be developed ever more rapidly. A broad range of purification tools has evolved to meet these needs but practical guidance concerning their use has lagged.

This book fills that void. It provides a major resource to the downstream processing community by integrating comprehensive information on purification tools with the specific features and requirements of monoclonal antibodies. Knowledge based on years of hands-on experience provides process designers from both academia and industry with valuable insights that will expedite development and assure them that they have have taken full advantage of the best that purification technology has to offer.

Gail K. Sofer,
Director of International Validation
Pharmacia Biotech, Inc.

Preface

"There is more to learn by climbing the same mountain a hundred times, than by climbing a hundred different mountains."
—Richard Nelson

The main reason I wrote this book is that I find monoclonal purification utterly fascintating, and I wanted to share that fascination in a way that would let others enjoy it as much as I. Certainly, the more enjoyable you find a task, the more likely you are to be successful. My hope is to facilitate that success by removing the obstacles that commonly prevent people from fulfilling their monoclonal purification goals.

To identify those obstacles, I relied on the purification questions my clients have asked me over the past decade. What are the options? How does a particular method work? What are its strengths? What are its weaknesses? How do I exploit it to its greatest benefit while minimizing the influence of its limitations? How does it fit with other methods? Where can I find more information?" Every aspect of this book has been designed to answer those questions.

Beyond content and organization, people requested a practical hands-on approach; one that integrates the unique characteristics of monoclonal antibodies with the specific capabilities and limitations of the various purification methods. There is an immense difference, for example, between ion exchange chromatography in general and ion exchange as practiced within the molecular and regulatory constraints of monoclonal purification. There are opportunities and obstacles that don't apply to any other group of proteins. This applies to every technique.

In the same spirit, it was frequently requested that examples used to illustrate key concepts derive as much as possible from industrial applications. I consider this to be one of the book's strongest features, but it involved a compromise. Much of the information is proprietary and permission to use the examples was contingent on omitting sensitive details. However enough information is included to let you judge for yourself if an example applies to your specific circumstances, and enough so that you can adapt it.

Purification-product names are omitted throughout the text for another reason. I wanted to avoid inadvertent bias favoring or disfavoring any particular supplier, and especially to avoid any implication that a certain set of results require the use of a specific product. Assume that the examples are representative of all closely related products. Where a specific product is needed, it is identified. Where a specific product class is required, I identified every product I was aware of.

I was also asked to keep technical explanations as intuitive as possible. Equations have a unique ability to integrate the influence of multiple parameters into a single expression, but they can be difficult to follow. Narrative suffers from linearity but lends itself to expressing ideas in more familiar terms. I chose the latter. This will hopefully provide a larger group of practitioners with an understanding that they can use to meet challenges creatively instead of reflexively.

Numerous people contributed to the development of this book in other ways. At the top of the list, Deb Neff and all the folks at Becton Dickinson Immunocytometry Systems: Noel Warner, Vernon Oi, Ken Davis, Neal Weinstein, Bill Godfrey, Don Ladd, George Herrel, Joe Link, the entire Product Development, and Manufacturing staffs. I can never thank them enough for their indulgence in providing me unlimited access to antibodies and laboratory facilities to develop ideas into practical methodology.

Many thanks to Gail Sofer, Jan-Christer Janson, Lars Hagel, and Makonnen Belew; some of my oldest and most valued friends in the biotech business. Also to Duncan Low, Peter Moore, Tim Hooper, Les Beadling, Anne Barry, Al Williams, Lars-Johan Larsson, Torgny Lindbäck, Eric Grund, Ursula Snow and many other good friends at Pharmacia. Likewise to Jerry

Volenec at Cytel, Peter Cartier at Rohm and Haas, Brandon Price at Microbiological Associates, David Buck and Diether Rechtenwald at AmCell, Paul Smith at BioAffinity Systems, Dennis Arvanitis at Dow Chemical, Scott Fulton and Dan Freymeyer at PerSeptive Biosystems, Peter Grandics and Bruce Hoffman at Sterogene Bioseparations, Joel Henner at Connective Therapeutics, Anne Moschella at Bioprocessing Ltd., Joe Machamer at Bio-Rad, and Andrea Knight at Genetics Institute. Their many contributions over the years have been much appreciated.

Finally, I offer my deepest gratitude to my clients. It's been the experience provided by working with them and hundreds of their products that has given me the orbital perspective necessary to write this book. I now turn this resource over to you, and I look forward to seeing where you take the field from here.

PSG
Tucson, Arizona
September 27, 1996

Table of Contents

Chapter 1

Introduction

"Order and simplification are the first steps toward mastery of a subject"
—Thomas Mann

What this book is about

The purpose of this book is to provide you with a guide to developing monoclonal antibody purification procedures that meet the requirements of both research and commercial applications. It is based on successful purifications developed for over 250 monoclonal-based products, addressing a wide range of diagnostic and therapeutic applications. It is supported by nearly 1000 citations from the scientific literature and enriched by the insights of skilled practitioners from throughout the industry. It incorporates over 100 figures and tables to illustrate key concepts.

The book's content and organization are based on 3 premises. The first is that every monoclonal antibody is unique. Each has a distinctive distribution of hydrophobic, positive and negative charge characteristics. In addition to immunospecificity, these features confer important secondary attributes like titer, stability and pharmacokinetic behavior. They also define the relationship of each monoclonal to the contaminants in its production medium, the diversity of which compound the fundamental uniqueness of each antibody. A clone's individuality is compounded further by its product application requirements: differences in purity and integrity; differences in regulatory requirements; differences in economic constraints.

The second premise is that every purification tool is unique. Each offers a distinctive set of capabilities and limitations. Since every antibody is different, and since its contaminants and application requirements are different, the strengths and weaknesses of the various purification techniques are likewise relative.

The third premise is that to consistently develop purifications that fulfill all of a product's requirements you must have a complete set of tools, you must know their relative strengths and limitations, and you must be able to adapt them to the needs of your situation.

How this book is organized

Purification methods are grouped according to their operational characteristics: precipitation, nonaffinity chromatography, and affinity chromatography. Each chapter begins with discussion of the mechanism by which the method operates. The intent is to provide an understanding of the technique, rather than just a description of process mechanics, so that you can fully appreciate its capabilities and limitations. Explanations are expressed as intuitively as possible and without equations, but they are rooted in physical chemistry. They incorporate the most recent information available. Consequently, they differ from explanations in many older purification guides.

Chapters continue with discussion of each method's attributes in the specific context of monoclonal purification. This will give you a good idea of the best you can expect from a method, and the features you might want to combine with other methods to construct a better overall process. This section also identifies circumstances under which a method's strengths may be compromised or even turned to disadvantage.

A corresponding section on limitations discusses typical compromises associated with each method. Points include identification of manufacturing logistical limitations, chemical incompatibilities, toxicity, and other issues of potential regulatory concern. Emphasis is on problem prevention but the discussion acknowledges that you don't always have the luxury of choosing your battles. Detailed suggestions are given for minimizing limitations you can't avoid.

"Method development" covers specific points that should be addressed to achieve robust purification procedures. Although the recommendations have been extensively field-tested, it doesn't mean that they represent the only way to develop purification methods. Furthermore, they make no distinction whether you are developing a product for in vitro or in vivo application. They provide a thorough framework but they rely on your knowledge of your own product requirements to decide what is applicable and to what degree.

A concluding section discusses how each purification method relates to others. It identifies notably productive and nonproductive combinations with other methods. The chapters close with suggestions for readings that provide good points of entry into supporting segments of the scientific literature. Extensive referencing is intended to facilitate deeper exploration.

Appendix I provides a series of foundation protocols. Discussion includes identification of the antibody classes and subclasses to which each protocol is suited, identification of the product applications to which it is most suited, materials and methods, and suggestions for refinements and variations. Each can be applied without modification or optimized according to your needs.

Appendix II addresses sample preparation. It describes a field-proven method for foulant removal that can be incorporated smoothly into almost any antibody purification scheme.

In practice

The scope and content of this book can give you the impression that monoclonal purification is a complicated difficult task. That's generally not the case and it's not the intent of this book to create that impression. On the other hand, the industry is changing. Pressure to control costs is growing, especially in the face of market competition. This has always been a concern with in vitro diagnostics. Now, with patent expiration looming for many first-generation products, it's becoming a factor for vivo products as well. At the same time, applications for new products continue to diversify, performance requirements grow increasingly stringent, and regulatory agencies are becoming more sophisticated in their evaluations.

These trends are not going to reverse. The needs of many products will continue to be fulfilled by a small subset of methods, but industry-wide, it's clear that demands placed on the next generation of products will require a more flexible selection of purification tools than ever before. The only way to meet these needs is to acknowledge and embrace them.

In spite of its scope and content, this book can't tell you the best way to purify any particular antibody. That's a journey you have to undertake for yourself. With experience you'll fall into the rhythm of the terrain, but it's infinitely variable and there is always something new over the next crest. When you encounter an obstacle—and you will—try a different approach. The only truly sustainable competitive advantage is the ability to learn faster.

Chapter 2

Precipitation Methods

"No idea is so antiquated that it was not once modern..."
—Ellen Glasgow

Fractionation of antibodies by precipitation has been in practice for nearly a century and a half.[1] By the time monoclonals emerged, a suite of precipitation mechanisms were already in wide commercial use with polyclonals. They included precipitation with inorganic salts, organic polymers, organic solvents; by electrolyte depletion; and by complexation with organic acids and bases.[2,3] Most of these methods have been applied successfully to purification of monoclonal antibodies.

Precipitation methods can be classified in 2 groupings; those that selectively precipitate antibodies, chiefly by altering the properties of the solvent; and those that selectively precipitate contaminants by complexation with agents that reduce their solubility. The first group includes precipitation with salts such as ammonium sulfate, with organic polymers such as polyethylene glycol (PEG), and by electrolyte depletion. The second group includes precipitation by organic complexants such as short chain fatty acids and organic bases. In addition to their unique selectivities, each method offers a different mix of practical advantages and limitations.

Precipitation with inorganic salts

Salt is the most common vector for monoclonal precipitation. The detailed mechanics by which precipitation occurs are a subject of continuing refinement but the basic elements are largely established. Precipitating salts are excluded from protein surfaces, leaving them preferentially hydrated.[4,5] With increasing salt concentration, it becomes energetically more favorable for proteins to form associations so they can share hydra-

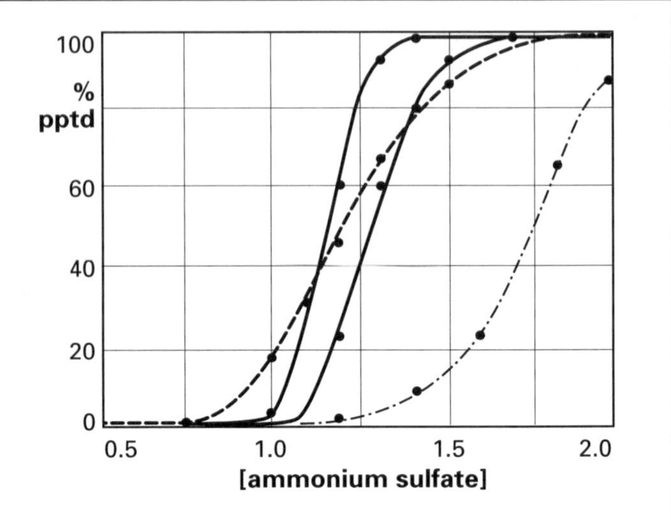

Figure 2.1 Ammonium sulfate precipitation of monoclonal and polyclonal IgG. The solid curves represent 2 different mouse IgG_1 monoclonals. The overlapping dashed curve is polyconal IgG. The light dashed curve to the right is albumin. The shallow slope of polyclonal antibodies reflects their heterogeneous origin, as well as the presence of aggregates and fragments. All results were obtained at pH 7.0.

tion shells rather than remain individually soluble. The high salt concentration concurrently suppresses long-range charge interactions. Hydrophobic surfaces associate with other such surfaces, facilitating aggregation, and eventually leading to precipitation.

Protein solubilities decrease exponentially with increasing salt concentration, yielding straight lines on semi-log plots.[6-8] Slopes and intercepts depend on the polar/nonpolar composition of a protein's surface and the properties of the bulk solution. Antibodies, being relatively hydrophobic, are among the first proteins to precipitate. Monoclonals in particular exhibit characteristically steep response curves due to their homogeneity (Figure 2.1). This translates into narrowly defined precipitation "zones" that facilitate separation from most of their more soluble contaminants.

effects of different salts

Different salts yield different selectivities. This was first observed by Hofmeister in his still-classic 1888 publication on the effects of various salts on protein solubility.[9] The ability of a given salt to promote precipitation correlates with the additive rankings of its component ions in the lyotropic series (Table 2.1).[4,5,10] Ions high in the lyotropic series are strongly excluded from protein surfaces. This makes them effective promoters of cohydrative protein association. Ions low in

Table 2.1. The Hofmeister series of lyotropic and chaotropic ions. The lyotropic effect is sometimes referred to as the "salting-out" effect, while the chaotropic effect is referred to as "salting-in".

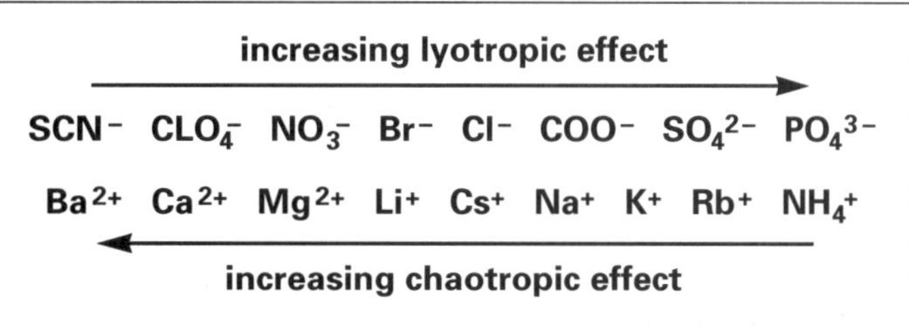

increasing lyotropic effect →

SCN^- ClO_4^- NO_3^- Br^- Cl^- COO^- SO_4^{2-} PO_4^{3-}

Ba^{2+} Ca^{2+} Mg^{2+} Li^+ Cs^+ Na^+ K^+ Rb^+ NH_4^+

← increasing chaotropic effect

the series are able to penetrate protein hydration shells and bind directly to their surfaces. This disrupts cohydrative association, favoring protein solubilization.

Older publications note a correlation between the surface tension increment of a given salt and its precipitating ability.[11,12] However, the deviations are substantial, and protein solubility behavior is accounted for more accurately by the ion exclusion/binding model.[4,5] Protein conformational changes due to allosteric interactions with specific ions cause selectivity variations that defy both models.[13-15]

While selectivity differences among salts can be substantial, only a few merit routine screening as part of method development. Besides being less efficient promoters of hydrophobic interactions, ions that bind to protein surfaces have a destabilizing influence.[4,5] Process choices are therefore best restricted to salts with both component ions high in the lyotropic series.[8]

pH effects

Selectivity is also influenced by pH, with proteins precipitating most readily at their isoelectric points (pI).[6-8,16] This indicates an electrostatic contribution that seems counterintuitive for a high salt environment. The rationale is that charged residues on protein surfaces lie within the protein's hydration shell. Short-range charge interactions are thereby shielded from the damping effects of excluded ions in the bulk solution. Moreover, the hydration water immediately surrounding charged residues is strongly electrostricted and not displaced by protein:protein associations. When a given polar residue is fully charged, it is maximally hydrated and depresses local hydrophobicity. When fully titrated, it is minimally hydrated and minimally influential. A protein's pI therefore reflects its point of minimal hydration, and maximum tendency to associate hydrophobically with other proteins.[17]

effects of protein concentration

Selectivity is strongly influenced by protein concentration.[6-8,16] Precipitation is most efficient at high protein concentrations. Salt concentration must be elevated to achieve the same selectivity for the same protein at a lower concentration. This highlights the important point that precipitation is an equilibrium process, which means that time is also a factor.

temperature effects

Precipitation efficiency declines with temperature.[6-18,16,18] This is a compound effect, reflecting kinetics, temperature dependent conformational changes of the protein, shifts in the pKs of polar residues, and temperature dependent changes in the relative solubility of the precipitating salt. An additional factor with antibodies is that some are cryoglobulins: progressively insoluble below 37°C. This is observed with 10–20% of IgMs and perhaps 1–2% of IgGs.[19-23] Cryoprecipitation is essentially independent of pH between 5 and 10, and only mildly suppressed by elevated salt concentrations. This behavior is consistent with most of the binding energy being derived from hydrogen bonding.[19]

attributes

Simplicity is the most attractive aspect of salt precipitation. The skills required to perform the technique are rudimentary. There are no hazardous chemicals. There is no requirement for specialized equipment.

purification performance

Salt precipitation delivers reasonably good purification for the effort, ranging from 50–80%, averaging around 60%. Consistently achieving the higher end of the range requires systematic method development and precise control of separation conditions. It also requires that a preliminary cut be conducted to remove contaminants more hydrophobic than the antibody, and that the precipitate be washed by resuspending it in clean precipitating buffer, then resedimenting it. This removes soluble contaminants that otherwise occupy the interstices of the precipitate. Preliminary cuts and washes tend to be unpopular because of the time and manual labor component, and because they reduce recovery, but if you are relying on salt precipitation as your primary purification method they may justify the sacrifice.

contaminants

Although salt precipitation is effective for removing the bulk of albumin, transferrin and other highly soluble contaminants, salt-fractionated monoclonals are invariably contaminated with polyclonal host antibody. Other major contaminants include lipoproteins, fibronectin, fibrinogen, α_2-macroglobulin, haptoglobin, β_1-haptoglobin, C1 and C3.[2,3] Its equilibrium mechanism makes it inefficient for removal of trace contaminants regardless of their solubility properties.

recovery of mass and activity

Mass recoveries range from 60–80%. Achieving the higher end usually requires sacrificing purity. This highlights one of the fundamental compromises with salt precipitation: you can't optimize both purity and mass recovery under the same conditions. Recovery of immunoreactivity per unit mass is usually quantitative.

limitations

Salt precipitation is portrayed as a mild chemical procedure, with little risk of denaturation or aggregate formation. This is consistent with the protein-stabilizing effects of lyotropic salts but contrary to many reports.[2-8,24-26] The discrepancy derives mostly from poor technique. Direct addition of dry salt causes denaturation of proteins at the hydrophobic interface of the dissolving crystals. Inadvertent foaming of the sample during resuspension has the same effect. The precipitating salt should be added as an aqueous concentrate, and care taken to minimize introduction of air during resuspension.[8] Turbid resuspensions are a sign of aggregates.

Another route by which denaturation and aggregation occur is by conducting ammonium sulfate precipitations at alkaline pH. Titration liberates strongly basic free ammonia, which denatures proteins—including antibodies.[3,27,28] If you wish to explore selectivity at alkaline pH, substitute potassium phosphate. It yields nearly the same average molar selectivity.[17]

Precipitates should never be resuspended in water. The lack of buffer capacity risks exposure of antibody to low pH and conductivity that can denature many antibodies, especially IgMs. Use a pH-neutral buffered solution for IgG. Include at least 0.05M sodium chloride for IgMs. Twice that may be necessary in some cases. Glycine is solubilizing and stabilizing for antibodies, and may

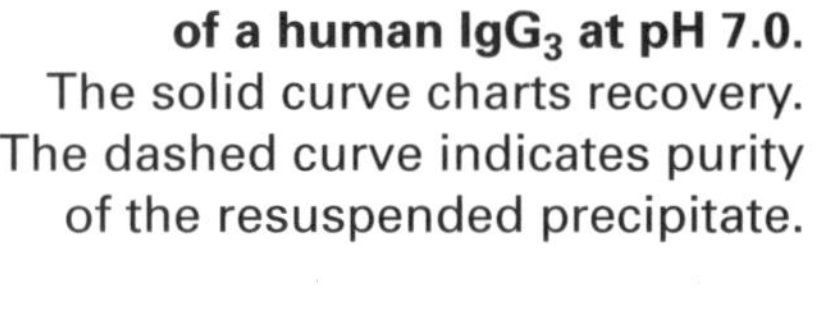

Figure 2.2 Recovery versus purity for unoptimized salt precipitation of a human IgG_3 at pH 7.0. The solid curve charts recovery. The dashed curve indicates purity of the resuspended precipitate.

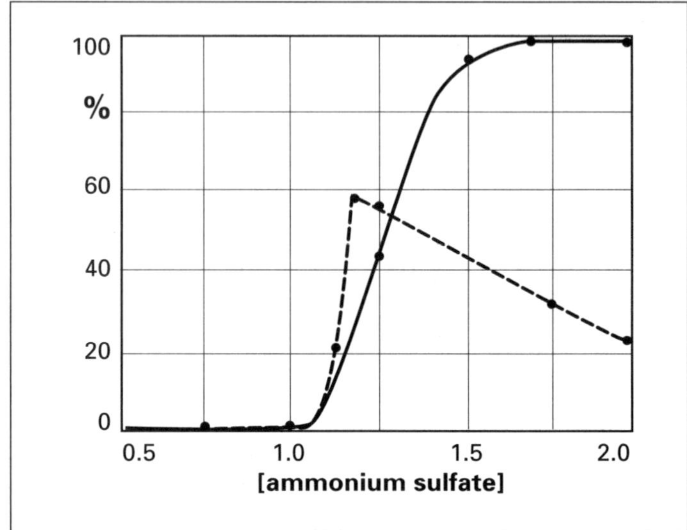

improve recovery of native product.[5,29-31] It is zwitterionic and won't interfere with downstream charge-based separation methods. However, it strips nickel from immobilized metal affinity (IMAC) columns.[32]

salt interference with downstream methods

Residual salt in the resuspended precipitate can interfere with downstream charge-based separation processes, such as ion exchange (IEC) and hydroxyapatite chromatography (HAC). Ammonium ions have chelating capacity and will interfere with IMAC.[32]

interference with electrophoresis

High salt interferes with electrophoretic assays by dehydrating the gel matrix and causing convective heating at the point of sample application. This creates banding artifacts. Generated heat can be sufficient to melt through the plastic backing on IEF gels and short circuit the system. This is usually not a problem with resuspended precipitates, but it's an important point to be aware of when analyzing supernatants for the presence of antibody, for example during method optimization or troubleshooting. A multisample microdialyzer ensures uniform electrophoretic conditions for analysis.

interference with protein modification

Ammonium ions interfere with many of the techniques used for protein immobilization or conjugation, and with total nitrogen assays.[8] In either case, precipitating with potassium phosphate or performing an intermediate buffer exchange step solves the problem.

Table 2.2. Performance and reproducibility of salt precipitation versus HIC for the same antibody.

	Precipitation		HIC	
	purity	yield	purity	yield
Lot# **1**	61%	65%	77%	86%
2	64%	65%	74%	88%
3	59%	65%	75%	85%
4	61%	65%	79%	84%
5	68%	65%	76%	86%
6	60%	65%	76%	88%
7	65%	65%	75%	83%
8	62%	65%	74%	86%
9	60%	65%	77%	87%
10	63%	65%	—	—
mean	**62%**	**64%**	**76%**	**86%**
range	9%	16%	5%	5%
S.D.	2.8%	5.6%	1.6%	1.7%

poor reproducibility

Reproducibility of salt precipitation processes is characteristically poor. Figure 2.2 illustrates the basis of the problem; the purity and recovery curves are offset from one another. Table 2.2 presents reproducibility data from an industrial precipitation process conducted strictly according to SOP. Comparison with a hydrophobic interaction chromatography (HIC) process for the same antibody highlights the problem.

aberrant precipitation variants

Be wary of some IgMs that produce clear precipitates. This appears to correlate with the tendency of many IgMs to cryoprecipitate. Such precipitates are often difficult to resuspend. The most effective approach is to include a strong hydrogen donor in the resuspension buffer, for example 1.0M urea.[19,33-35] Risk of denaturation at this concentration is nil. Another advantage with urea is that it does not interfere with charge-based downstream separation methods. Glycine may also be helpful. Increasing salt concentration generally does not provide significant benefit, but it can be increased up to 2.0M sodium chloride as a last resort. Warming the buffer may facilitate resuspension but elevates the risk of proteolysis.

A hidden liability with clear supernatants is the ease of accidentally discarding product. The risk is highest with samples that have been stored in the cold. The antibody precipitates on the surface of the container, where it is easily overlooked. If you encounter an antibody that gives clear precipitates, evaluate temperature-dependent solubility thoroughly. Make sure that manufacturing documents include adequate warnings about bringing all materials to temperature before taking test samples, transferring product between containers, or processing.

buffer incompatibilities

A minority of strongly basic antibodies are able to form stable crosslinks with polyvalent anionic buffers, such as phosphate and sulfate. This particularly includes IgG_3s. On a gross level these antibodies resuspend without undue difficulty but they are persistently turbid and product is lost steadily over time. Sodium or potassium fluoride in the resuspension buffer will often correct the problem. Experimentation is required to determine the effective concentration.

problems with metal contamination

It is important to use high grade salt preparations to minimize metal contamination. Divalent metal cations bind directly to protein surfaces, reducing their relative hydration, altering their net charge, and reducing their pIs.[3-5,36] All of these factors alter selectivity. Metal binding also compromises product stability. These effects make metal control a key determinant of product quality and process reproducibility.

Even low part-per-million (ppm) metal concentrations are significant. For example, 1ppm nickel in a 1mg/mL solution of IgG corresponds to a molar excess of >50 fold. EDTA will prevent formation of coordination complexes between metal ions and protein-polycarboxy sites, but it can't prevent their complexation with histidyl residues.[32,37] This is a particular concern with IgGs because of a highly conserved histidyl cluster at the juncture of the $C\gamma2$ and $C\gamma3$ domains.[37-39] Imidazole blocks this interaction, as do histidine and histamine.[32] Although including imidazole and EDTA in precipitating buffers will minimize the influence of lot-to-lot variations in metal contamination, the best place to control variation is at the source. Use American Chemical Society (ACS) grade salts—or better. Log certificates of analysis to document long-term material quality profiles.

Method development

Salt precipitation processes that deliver the performance and reproducibility required to support commercial applications require systematic development for each monoclonal. The key variables are the salt concentration, antibody concentration of the starting solution, pH, temperature, time, and the method by which the precipitating salt is introduced. Resuspension buffer and technique are also important.

protein concentration

The first step is to lock in the antibody concentration of the starting material. Since precipitation is an equilibrium process you'll want the starting concentration as high as possible. However, product concentration often varies among production lots. To obtain good reproducibility, you need to specify the lowest concentration of antibody that Manufacturing will receive, and dilute more concentrated samples to that level. Consult with your production staff to determine the range of product concentrations Manufacturing is likely to experience.

process temperature

Temperature is not a practical tool for manipulating selectivity over wide ranges. The manufacturing logistics are prohibitive. Choose either the ambient or cold-room temperature of your manufacturing facility and concentrate on maintaining tight control. Cold processing may be preferable if proteolysis is a concern. Protein conformation is also more stable in the cold. However, cold processing can be a burden on technicians. It's important to weigh its practical value against the convenience of room-temperature processing.

operating pH

Set the pH of the separation to match the pI of the antibody. If the pI is alkaline, use potassium phosphate. Otherwise, evaluate ammonium sulfate, potassium sulfate, or both. Prepare a series of solutions that vary by increments of 0.1M salt from 1.5–2.0M, all with the same antibody concentration. Incubate 30 minutes then centrifuge down the precipitate at the process temperature. Decant and retain the supernatant and resuspend the precipitate.

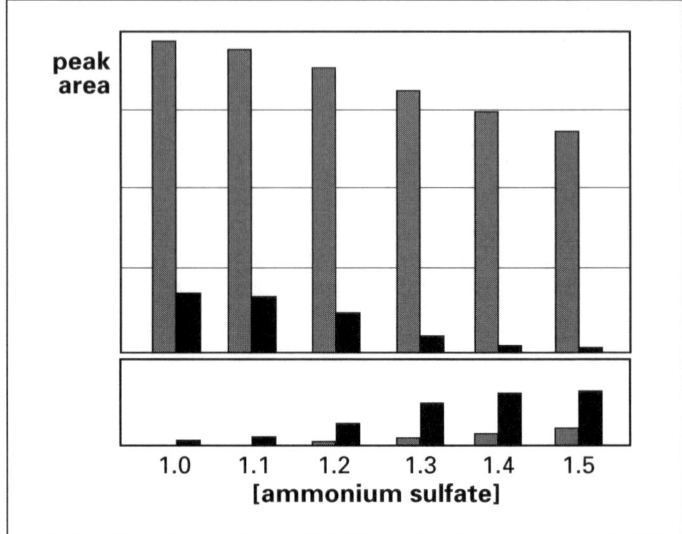

Figure 2.3. Affinity chromatography evaluation of ammonium sulfate precipitation conditions. The gray bars indicate the area of the unretained peak. Black bars indicate the area of the eluted peak. The upper block illustrates results from supernatants at the indicated salt concentrations. The lower block illustrates results from the resuspended preicpitates.

evaluating preliminary results

Affinity chromatography is a convenient method for measuring antibody content of both the resuspended precipitate and the supernatant. Protein A and protein G can be used for IgG monoclonals, while anti-light chain ligands can be used for other classes.[40-43] Comparing the fall-through fraction with the eluting fraction provides an estimate of relative purity and recovery (Figure 2.3). Most affinity methods are not affected adversely by the high salt content of supernatant.

incubation time and method of salt addition

The last step is to set the incubation time. This should be done with the salt addition method that will be used in the actual manufacturing process. Some version of the following method is recommended. Set the raw sample stirring and add liquid salt concentrate through a peristaltic pump so that the addition occurs over a period of at least 30 minutes. This will yield a much finer homogenous precipitate that will resuspend more easily with fewer aggregates. Continue to incubate at time increments of your choice. If you are processing at 4°C, consider overnight (16 hour) incubation. Duplicate the experiment at 0.05M and 0.10M salt below target. This will provide an objective basis for adjusting salt concentration to compensate for the increased precipitation from the longer incubation time.

Figure 2.4 illustrates the purity and recovery curves

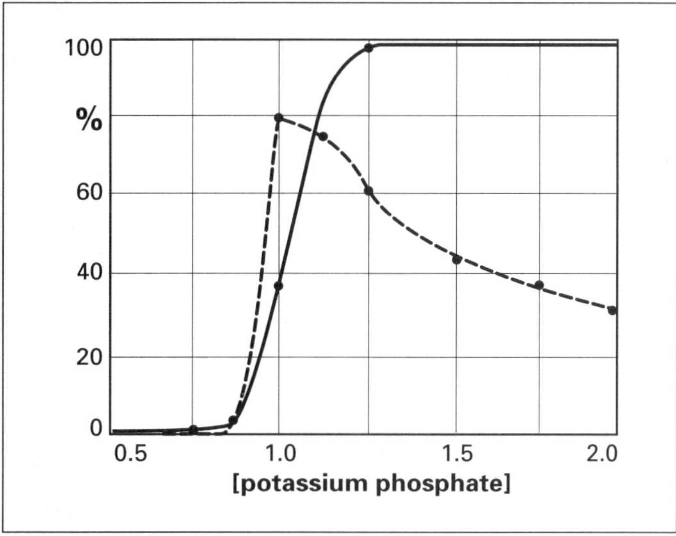

Figure 2.4. Recovery versus purity for optimized salt precipitation of human IgG_3. This series of experiments was conducted at pH 8.3, the same as the pI of the antibody. The solid curve charts recovery. The dashed curve indicates purity of the resuspended precipitate. Compare these profiles with the same antibody in Figure 2.2.

for a human serum myeloma IgG_3 after optimization. Compare this with the unoptimized curves for the same antibody in Figure 2.2. Purity and recovery both improved nearly 20%.

Precipitation with polyethylene glycol

Polyethylene glycol (PEG) is a nonionic polymer of ethylene glycol. It has been used for industrial preparation of polyclonal antibodies for decades.[2,3,44-47] In parallel with its polyclonal applications, it's applied more frequently to monoclonal IgMs than to IgGs. The mechanism by which precipitation occurs is related to that of salt precipitation but with critical distinctions that translate into unique selectivity.

mechanism

PEG is excluded from protein surfaces, resulting in their preferential hydration.[48-50] As PEG concentration increases, it becomes energetically more favorable for proteins to associate and share their hydration shells rather than remain individually soluble. In contrast to salt precipitation, long range charge interactions are fully operative and more influential on selectivity. As with salts, protein solubility decreases exponentially with increasing PEG concentration.[51-53]

Proteins of similar size precipitate according to their relative solubility—least soluble first—while proteins of similar solubility fractionate according to their relative size—largest first.[53-55] Antibodies fit this pattern.

Figure 2.5. PEG precipitation of polyclonal antibodies. Human serum samples were diluted 1:1 with a 2x concentrate of PEG-6000 in 0.10M Tris, 0.15M sodium chloride, pH 8.0. Data replotted from reference 56.

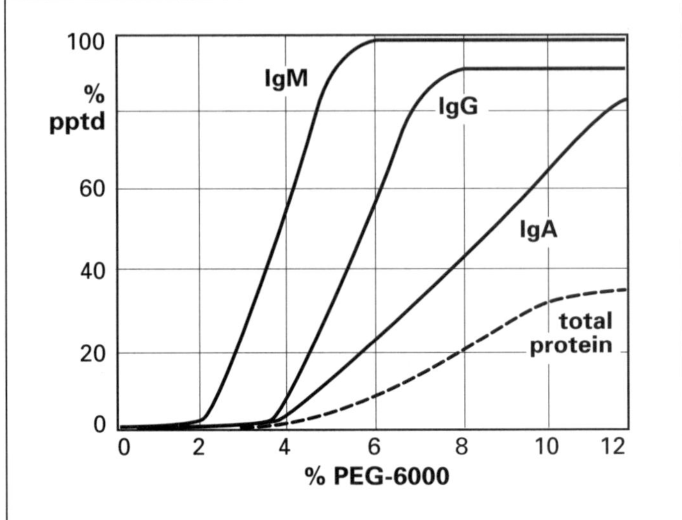

IgMs precipitate more readily than IgGs even though their precipitation characteristics in ammonium sulfate are relatively similar. IgGs precipitate more readily than IgAs, reflecting the higher solubility of the latter (Figure 2.5).[2,3,56]

The dominant process variables in PEG precipitation are the size and concentration of the polymer. PEG with an average molecular weight of ~6000 is used most commonly.[57] Smaller polymers like PEG-400 have been reported to improve resolution by as much as 2-fold.[58] PEG-400 is also easier to use since it's a liquid and has only ~1% the molar viscosity of PEG-6000.[59] However, the net viscosity reduction is not as substantial as this implies, since a much higher concentration is required to achieve the same result.[57,60] Polymers larger than Mr 6000 appear to offer no advantage.[61] PEG precipitation is generally less responsive to temperature variation than precipitation with salts.[52,54,62]

attributes

The application strengths of PEG precipitation are largely the same as for salt precipitation. It's a simple technique with no requirement for hazardous chemicals or specialized equipment. It supports reasonably good purification and recovery, and its stabilizing effect on protein conformation makes it well suited to fractionation of labile proteins.[21,61,63-66]

purification performance

While its selectivity is different from salt precipitation, and much more responsive to variations in pH, its average purification capability is about the same: 50-80%. The bulk of albumin, transferrin and other smaller more soluble contaminants are removed easily. The major contaminants are nonspecific immunoglobulins, lipoproteins, fibronectin, fibrinogen, α_2-macroglobulin, haptoglobin, β_1-haptoglobin, C1q and C3.[2,3]

compatibility with downstream methods

PEG's nonionic composition provides an important advantage over salt precipitation: it doesn't interfere with charge-based downstream separation methods. Even high-PEG-supernatants can be applied directly to IEC, HAC, or affinity supports without depressing binding efficiency.[60,67] Since PEG itself does not adsorb to any of these media, they all provide effective means for its removal.

limitations

PEG's weaknesses are similar to those of salt precipitation. Either purity or recovery can be enhanced with method development, but one is achieved at the expense of the other. Due to its reliance on an equilibrium mechanism, it suffers the same inefficiency for removal of low level contaminants, compounded by retention of supernatant contaminants in the interstices of the precipitate.

interference with assays

High PEG concentrations interfere with electrophoretic separations by dehydration of the gel at the point of sample application. They also interfere with analytical size exclusion chromatography (SEC) separations. PEG is preferentially excluded from chromatography media surfaces just as it is from proteins.[48] In the presence of high PEG concentrations, proteins share their hydration shells with the chromatography media and are either retarded or retained.[56,67-72] Since PEG-6000 behaves hydrodynamically like a charge-neutral globular protein with a molecular weight of 50–100kD, the effect migrates down the column with the proteins. This causes aberrant elution profiles.

high viscosity

PEG's high viscosity increases backpressure and may require reduction of flow rate during SEC. It also induces viscous fingering. This refers to nonideal flow distribution that occurs as a result of a low viscosity solvent punching holes through a higher viscosity sample layer.[73-75] Fingering detracts from resolution and makes SEC unsuitable for removing high levels of PEG-6000 from supernatants. PEG-400, which behaves hydrodynamically like a globular protein of about 5-10kD, does permit removal by size-based methods but viscosity remains an issue.

PEG removal

A nonchromatographic approach for removing high concentrations of PEG from supernatants is to add 4.0*M* sodium chloride to a final concentration of 2.0*M*. This causes the PEG to "precipitate" and float to the surface. About 95% of the protein remains in the aqueous phase. The PEG can then be decanted, and the remaining protein solution desalted by the method of choice.[76] A variation of this technique is to precipitate the proteins from the PEG with ammonium sulfate. The biphasic supernatant is discarded after centrifugation and the precipitate resuspended as usual.[77]

aberrant precipitation behavior

The precautions with cryoglobulins discussed under salt precipitation apply equally to PEG precipitation; likewise crosslinking of basic antibodies with phosphate.

method development

The first choice in method development is the polymer molecular weight. PEG-6000 provides the advantage of having a body of reference data for antibody purification, but the disadvantage of size-based methods being ineffective for its removal. Lower molecular weights should provide better resolution, be removable by size-based methods, and provide the convenience of already being in liquid form, but you'll have to establish the working ranges yourself. The following discussion is based on PEG-6000.

working ranges

The logistics of method development are the same as with salt precipitation. Prepare a stock solution of 25% PEG-6000 at the antibody's pI. Set process temperature and antibody concentration of the starting material. For IgGs, evaluate PEG-6000 in increments of 1% from 6–10%; for IgMs, 4–8%. PEG interference with electrophoretic separations makes affinity chromatography the most convenient analytical option. Finally, set the incubation time, adjusting PEG concentration as necessary. As with salt fractionation, better results are obtained with gradual PEG addition.

resuspension buffers

Resuspend precipitates in a buffer that contains at least 0.05M sodium chloride to prevent euglobulin precipitation. This may leave you having to redilute the sample to load it onto an ion exchange column, but that's a better compromise than risking precipitation and possible denaturation of your product. As your overall process takes shape, you can reformulate the resuspension buffer to maintain product solubility, but bring it into closer conformance with sample application requirements for downstream purification steps.

Precipitation by electrolyte depeletion

Precipitation by electrolyte depletion is traditionally referred to as euglobulin precipitation. The technique dates from the earliest classifications of serum proteins, the euglobulins being those that precipitated in water.[1] The major class of antibodies precipitating under these conditions is IgM, and it's the only class to which the technique is routinely applied.

mechanism

As ions are depleted from bulk solution, protein-protein electrostatic interactions replace the protein-ion interactions that take place in physiological environments. Since most IgMs carry significant complements of both negative and positive charges, and are poorly soluble to begin with, their self-associative tendencies are enhanced. The charge pairs that bind the complexes are electrostatically neutralized, reducing net charge of the complex and further depressing solubility. Although the initial attraction among coprecipitants is electrostatic, their associations may be stabilized by hydrogen bonding and hydrophobic interactions.

attributes

Euglobulin precipitation provides a unique selectivity but its performance is highly variable. Purity is typically inferior to other precipitation methods, occasionally as high as 75%, but often less than 50%. The IgM fraction contains IgG, ceruloplasmin, β_{1c}-globulin, low density lipoproteins, α_2-macroglobulin, and many less well characterized contaminants.[3] It is important to appreciate that coprecipitation of a given contaminant does not necessarily indicate that it too is a euglobulin. Its presence may simply reflect electrostatic complexation with the product.[78-80] This accounts for the frequent presence of DNA in euglobulin precipitates.

limitations

Mass recovery can be as high as 80% but tends to be much lower. Cloudy resuspensions indicate persistent aggregates and often warn of permanent denaturation. Activity recovery also tends to be depressed, apparently reflecting global changes in antibody conformation. This is indicated by increased susceptibility to proteolysis and changes in the ability of the antibody to be bound by affinity ligands.[81,82] Mass and activity recoveries are best when precipitation is conducted at moderate pH: 6.0–8.0. Precipitation below pH 5.0 often results in abysmal mass recoveries and grossly altered antibody function.

This technique is adaptable to a chromatographic format: euglobulin adsorption chromatography, as described in chapter 8. Performance and scalability of the chromatographic format are so superior that there is little reason to pursue precipitation.

Precipitation with Octanoic acid

Precipitation with octanoic acid, also called caprylic acid, is an established method for commercial fractionation of various blood products (Figure 2.6).[2,3] It is one of several fatty acids that have been evaluated for antibody purification and the only one adapted to monoclonals.[83-94] Its mechanism of action is not addressed in publications describing its use, however physicochemical studies of interactions between similar compounds and protein surfaces provide the foundation for a working hypothesis.

mechanism

At low pH, the negative charge on the carboxyl moiety of octanoic acid is titrated.[95] The hydrophobicity of its octyl moiety then dominates its character. It penetrates protein hydration shells and binds to their surfaces, imparting an increase in hydrophobicity.[96] For acidic proteins, which are minimally soluble at low pH, the increased hydrophobicity results in their precipitation. Octanoic acid also binds to antibodies but because of their basic pIs they retain sufficient charge to counteract its desolubilizing effect, and they remain in the supernatant. After removal of the precipitated contaminants, increasing pH to neutrality restores the charge to the carboxyl moiety of the complexant, favoring its reassociation with the bulk solution and subse-

Figure 2.6. N-octanoic acid. In its titrated form (above) octanoic acid is roughly the hydrophobic equivalent of the octyl ligand used for hydrophobic interaction chromatography.

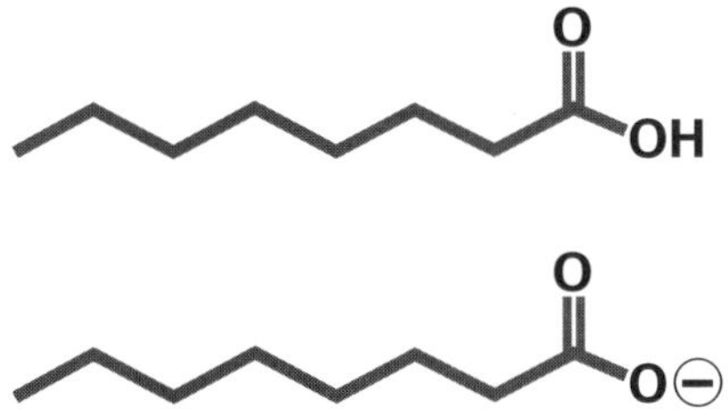

quent removal. Overall selectivity mimics cation exchange chromatography.

conductivity effects

Conductivity has a minor effect on selectivity in comparison to pH. Most publications recommend conditions with conductivity equivalent to ~0.05M sodium chloride. Increasing conductivity by addition of nonprecipitating salts has negligible effect. Addition of precipitating salts like 0.3–0.5M ammonium sulfate increase precipitation efficiency. Higher concentrations cause the antibody to precipitate with the contaminants.

temperature effects

Reliance on hydrophobic interactions makes performance temperature dependent. Purity diminishes with reducing temperature; recovery increases. For users preferring to conduct the process in the cold, the selectivity shift can be compensated by addition of precipitating salts to the reaction mixture, in the range indicated above.

kinetics and stoichiometry

Protein and complexant concentrations affect performance, but this seems to reflect stoichiometry rather than kinetics. The key factor is apparently the complexant:protein ratio. Ratios to either side of optimum result in reduced antibody purity. So long as the complexant:protein ratio is appropriate, absolute complexant concentrations appear not to be critical. Published reaction times average 20-30 minutes but equilibrium is actually achieved within ~10 minutes.[86]

attributes

Octanoic acid precipitation is simple, rapid, inexpensive, and applicable to all antibody classes from all species, including IgY from egg yolk. It can provide single-step total IgG purity greater than 90% and seldom provides less than 70%. In addition to nonspecific immunoglobulins, characteristic contaminants include ceruloplasmin, α_1-acid glycoprotein, prealbumin, and small amounts of albumin. Mass recoveries range from 50–90%, with basic IgGs at the top of the range, acidic antibodies at the bottom.

virus inactivation

The ability of octanoic acid to inactivate lipid-enveloped viruses is well documented.[95,97] This is not to suggest that the conditions under which antibodies are purified represent peak virucidal efficiency, but they may contribute to overall process clearance.

DNA and endotoxin removal

Octanoic acid precipitation should not be expected to support reduction of DNA contamination. Endotoxin clearance is undefined and unpredictable. On the one hand, the lipophilic character of the complexant should favor its association with endotoxin. On the other, they have the same net charge; mutual repellency may prevent the stable association required for precipitation.

nontoxicity

Octanoic is a naturally occurring fatty acid found in a variety of plant and animal fats. Low concentrations (~4mM) have been used to stabilize intravenous preparations of human albumin for nearly 50 years and it proved nontoxic when administered intravenously to mice.[98,99] One isotopic study indicated that the molar ratio of residual complexant to antibody was ~0.4/1.[86, 100] Secondary purification with ammonium sulfate precipitation reduced it to ~0.04/1.

limitations

Although residual complexant levels are low, they have been reported to invalidate hemagglutination assays.[87] Since most of the receptors on IgG are distinguished by elevated local hydrophobicity, this makes them prime binding sites for residual complexant.[101] The potential impact on pharmacokinetics and other aspects of antibody performance make it imperative to validate noninterference.

interference with electrophoresis

Residual octanoic acid increases anodal mobility of antibodies on native agarose gels.[102] No such effect has been reported for SDS gels. However, SDS and octanoic acid have the same net charge and differ only in hydrophobicity. SDS probably displaces residual complexant, but may simply mask its effects.

antibody solubility limitations

Antibody solubility limitations have a compound negative effect. Even in the absence of octanoic acid, many antibodies are barely soluble under the low pH, low ionic strength precipitation conditions. Such antibodies coprecipitate with the contaminants. When poorly soluble IgGs are secondarily purified by ammonium sulfate precipitation, product losses are far greater than encountered with ammonium sulfate alone. This is experienced as poor resuspendability of precipitate. It appears to result from the precipitating salt environment favoring secondary hydrophobic interactions between residual octanoic acid and the antibodies.The hazy resupended precipitates often contain high aggregate populations suggestive of denaturation.

reducing losses

These losses can be minimized by a combination of 2 treatments. First, increase the conductivity of the working octanoic acid solution by adding sodium chloride to a final concentration of ~0.1M. The solubilizing effect supports improved antibody recovery. The second step is to reduce residual octanoic acid content. Titrate the supernatant to pH 6.5–7.0, then add ~2% (w:v) of Dowex AG1X2 (chloride salt, 400 mesh) and stir 1–16 hours in the cold.[103,104] This resin is a hydrophobic strong anion exchanger, with a pore size distribution that admits only low molecular weight compounds. It scavenges free octanoic acid, maintaining the steepest possible concentration gradient, thereby encouraging dissociation of bound complexant from the antibody. This step should also be taken before any chromatographic separation to protect column media from fouling.

reagent quality and technique

The quality of the complexant is critical; obtain the highest grade reagent available. Technique is also important. Batch addition of complexant results in uncontrolled loss of antibody and poor purification. For small samples, for example 5mL in a 15mL conical bottom tube, add complexant gradually from a pippetting device while vortexing vigorously. Submerge the pipette tip in the solution during addition. For process scale addition, adapt the method described earlier for addition of ammonium sulfate. Poorly controlled addition of complexant is a frequent cause of scale-up problems.

odor

Octanoic acid has a disagreeable, strong, permeating, persistent odor reminiscent of an α-male goat.

method development

Publications have tended to recommend generic process conditions, but they have been directed mostly toward screening small volumes of large numbers of research samples, where strict process control and economics were not overriding issues. Commercial applications require individual optimization.

basic method

Dilute the raw sample to ~10mg/mL, using 0.05M sodium acetate, 0.05M sodium chloride, pH 4.0 as a diluent. Adjust the pH of 3 aliquots to 4.2, 4.5, and 4.8. Add 1µL of complexant per mg of protein (10µL/mL). Mix for 30 minutes at room temperature. Centrifuge and decant the supernatant, then screen it for antibody.

optimization

Select the highest pH that gives an acceptable balance of purity and recovery. If all pH values show significant albumin contamination, pick the best one and increase complexant concentration by 50%; again if necessary. Even if purity and recovery are excellent, try to reduce the raw material dilution factor while maintaining the complexant:protein ratio. The more you can reduce process volume, the more convenient you make scale-up. If you have to compromise, it's better to optimize recovery over purity. A second purification step will be required in any case, so there's no need to place a disproportionate burden on the octanoic acid step.

Precipitation with Ethacridine

Ethacridine (2-ethoxy-6,9-diaminoacridine lactate) is one of several organic bases used for commercial fractionation of plasma proteins, and the only one that has been adapted systematically for purification of monoclonal antibodies (Figure 2.7).[105-111] It is strongly hydrophobic, positively charged, and used at alkaline pH. This might lead you to believe that the precipitation mechanism—except for the charge reversal—is similar to octanoic acid. It's not. The complexant is applied at much lower ionic strength than octanoic acid and it precipitates contaminants by forming stable insoluble ionic complexes. Functionally, ethacridine operates like a liquid-phase anion exchanger, with selectivity to match.[105,106]

attributes

Ethacridine precipitation is simple, rapid, inexpensive, and applicable to all IgGs. It supports 60–80% purity. Supernatants are characteristically contaminated with

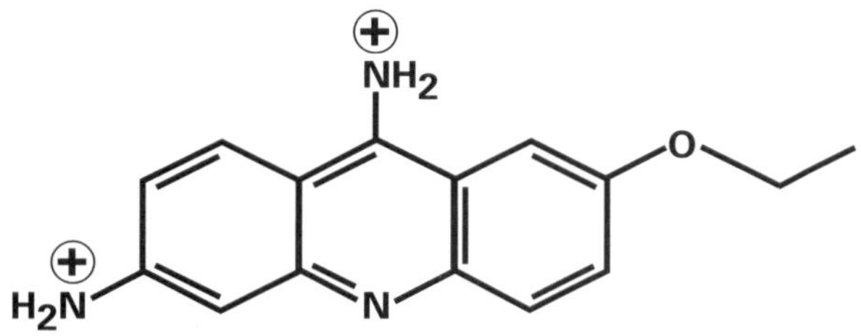

Figure 2.7. Ethacridine. Only the lactic acid salt is used for protein purification. It is marketed under a variety of tradenames, the most common of which is Rivanol (Hoechst).

transferrin, small amounts of α_1-acid glycoprotein, hemopexin, properdin factor B, cholinesterase, α_{2HS}-glycoprotein, nonspecific IgG and IgA.[2,3,107,108] Contamination with albumin indicates loss of pH control from liberation of ethacridine's lactic acid counterion during the reaction.[107,108]

recovery

Mass recoveries range from 60–90%. They are best with basic IgGs. Activity recoveries are typically quantitative. Negligible aggregate content is one of the method's hallmarks.[107,111,112]

removal of viruses, DNA and endotoxins

The ability of ethacridine to condition raw production media is at least as valuable as its ability to fractionate proteins. It is especially effective for removing aggregates, clots and cellular debris.[107] It precipitates lipids, including pristane, mineral oil, fatty acids, lipoproteins, phospholipids, endotoxins, nucleic acids, and viral particles.[106,110,113,114] It is also an effective bacteriocidal agent; translation: it's toxic.[3] Be sure to obtain material safety data sheets and adhere to all precautions.

Ethacridine's sample conditioning abilities are mediated primarily by electrostatic crosslinking and enhanced by its hydrophobicity, but its ability to intercalate is an important contributory factor with DNA. Its structural homology with the phenothiazine dyes suggests that electron transfer may be a factor as well. Photo-excitation of these dyes causes destruction of complexed viral lipid envelopes and nucleic acids, yielding more than 6 logs of viral inactivation.[115,116]

monitorability

Ethacridine's fluorescence provides a sensitive method for validation of its removal. The cohesiveness of ethacridine precipitates makes centrifugation unnecessary.

limitations

Ethacridine forms insoluble salts with halides and other inorganic anions.[111] This requires advance sample preparation to clear the salts. It may be tempting to simply add a compensatory excess of ethacridine, but the concomitant liberation of lactic acid exerts greater constraints on pH control. This doesn't mean that you can't do it; just that you need to be prepared.

buffer restrictions

Choice of process buffer is a challenge since neither the buffer nor the counterion can be inorganic anions. BICINE is a zwitterionic buffer with a pK of 8.04, ensuring good buffer capacity through the range from pH 7.5–8.5. It contributes practically nothing to conductivity and it's titrated to pH with sodium hydroxide, avoiding concerns about ethacridine coprecipitation with chloride ions.

useful only for IgGs

Recoveries of nonIgG antibodies are prohibitively poor with ethacridine purification. They can be improved by reducing pH, but at the direct expense of purity. Ordinarily this wouldn't be worthwhile, but given ethacridine's delipidating, anti-DNA, anti-endotoxin, and anti-viral properties, sacrificing purity in the initial process step may pay off in overall process efficiency.

removal of residual complexant

Studies have not addressed retention of residual ethacridine by purified antibodies, but it seems only prudent that affirmative steps be taken to remove it. The most direct approach is to selectivity precipitate it with inorganic anions.[107,111] Another is to incubate the purified antibody with a solid phase cation exchange medium. As with the corresponding method for removal of octanoic acid, the ideal exchanger is either nonporous or has a pore size distribution small enough to exclude the antibody. Reducing pH to ~6 increases the charge on the ethacridine and favors its dissociation.

UV interference

Ethacridine absorbs UV at 280nm, which can interfere with spectrophotometric monitoring. Its intense yellow color makes it messy to work with, but removes all ambiguity about where clean-up is required.

method development

Given consistent conductivity, selectivity is mainly a function of pH. Equilibrate 3 small aliquots of raw antibody to 0.05M BICINE; one at pH 7.50, one at 7.75, the other at pH 8.00. Prepare a stock solution of 0.2M ethacridine in 0.01M BICINE, pH 8.0. While vigorously vortexing, gradually add ethacridine stock to 10% v:v. A thick yellow precipitate will begin to form immediately. Incubate for 20 minutes at room temperature. Decant the supernatant and filter. Precipitate residual ethacridine from the filtrate by adding 0.2M MES, 4.0M sodium chloride, pH 6.0 (25% v:v). Incubate for 10 minutes then buffer exchange the samples on disposable columns and analyze them. Choose the pH that provides the best balance of purity and recovery.

If you plan to bypass the initial buffer exchange step, method development is more complicated and process variation is higher. You'll need to increase ethacridine concentration to adjust for coprecipitation with inorganic anions. Since anion coprecipitation will release a corresponding concentration of lactic acid, you'll have to increase the buffer concentration to compensate. Even though BICINE is zwitterionic, the sodium hydroxide required to titrate it will increase conductivity. Precipitation efficiency will be reduced.

Precipitation methods in perspective

Precipitation methods support a wider diversity of processing options than is generally realized. They offer selectivities mimicking the major chromatography methods. Salt precipitation parallels HIC (chapter 6). PEG precipitation parallels hydrophilic interaction chromatography (chapter 8). Euglobulin precipitation parallels euglobulin adsorption (chapter 8). Octanoic acid and ethacridine precipitation parallel cation and anion exchange (chapter 4).

A number of 2-step combinations provide adequate purity for in vitro diagnostic products without any requirement for chromatography columns or sophisticated equipment. The most publicized is octanoic acid followed by salt precipitation. Substituting PEG for salt produces equivalent results, as does substituting ethacridine for octanoic acid in either process.

Precipitation methods also make effective partnerships with chromatographic methods. Ethacridine with SEC is one such example. HIC following either ethacridine or octanoic acid works very well. Salt precipitation followed by IEC is the classical IgG purification method.

Recommended reading

For broad ranging discussion of precipitation in industrial preparation of therapeutic proteins see Methods of Plasma Protein Fractionation and The Molecular Biology of Human Proteins.[2,3]

For an introduction to the physical chemistry of interactions between protein surfaces and various classes of chemicals, consult references 4, 5 and 96. The reference by Dixon and Webb is considered the classic in salt precipitation.[8] For the mechanism of PEG precipitation see reference 48; for general application information consult reference 57. PEG precipitation of polyclonal and monoclonal antibodies is discussed in references 44–47 and 56. Begin with references 84-86 for discussion of octanoic acid precipitation. Reference 107 describes ethacridine precipitation specifically for purification of monoclonal antibodies.

References

1. P. Panum, 1852, *Virchov's Arch. Path. Anat.*, **4** 419
2. J. Curling, 1980, Methods of Plasma Protein Fractionation, Academic Press, London
3. H. Schultze and J. Heremans, 1966, Molecular Biology of Human Proteins, Vol. 1, Elsevier, New York
4. T. Arakawa and S. Timasheff, 1982, *Biochemistry*, **21** 6545
5. T. Arakawa and S. Timasheff, 1984, *Biochemistry*, **23** 5912
6. E. Cohn and J. Ferry, 1943, in Proteins, Peptides and Amino Acids as Ions and Dipolar Ions, (E. Cohn and J. Edsall, eds.) p. 586, Reinhold, New York
7. E. Cohn and J. Edsall, 1943, Proteins, Peptides, and Amino Acids as Ions and Dipolar Ions, Reinhold, New York
8. M. Dixon and E. Webb, 1961, *Adv. Protein Chem.*, **16** 197
9. F. Hofmeister, 1888, *Arch. Exp. Pathol. Pharmakol.*, **24** 247
10. V. Parsegan, 1995, *Nature*, **378** 335
11. W. Melander and C. Horvath, 1977, *Arch. Biochem. Biophys.*, **183** 200
12. B. Roetger et al, 1989, *Biotechnol. Progr.*, **5** 79
13. W. Melander et al, 1977, *Arch. Biochem. Biophys.*, **183** 200
14. J. Rosengren et al, 1975, *Biochim. Biophys. Acta*, **412** 51
15. H.-L. Wu et al, 1986, *J. Chromatogr.*, **371** 3
16. A. Green, 1931, *J. Biol. Chem.*, **93** 495
17. P. Gagnon and E. Grund, 1996, *BioPharm*, **9**(5) 54
18. J. Taylor, 1953, in The Proteins (H. Neurath and K. Bailey, eds.) Vol. 1, part A, p. 2, Academic Press, New York
19. C. Middaugh and G. Litman, 1977, *J. Biol. Chem.*, **252** 8002
20. G. Litman et al, 1981, *Immunol. Commun.*, **10** 707
21. C. Middaugh et al, 1980, *J. Biol. Chem.*, **255** 6532
22. R. Weber and L. Clem, 1981, *J. Immunol.*, **127** 300

23. C. Middaugh et al, 1978, *Proc. Nat. Acad. Sci. USA*, **75** 3440
24. S. Burchiel, 1986, *Met. Enzymol.*, **121** 596
25. J.-P. Bouvet et al, 1984, *J. Immunol. Met.*, **66** 299
26. A. Jehanli and D. Hough, 1981, *J. Immunol. Met.*, **44** 199
27. S. Hjerten et al, 1986, *J. Chromatogr.*, **359** 99
28. Y. Kato et al, 1984, *J. Chromatogr.*, **298** 407
29. J. Porath and N. Ui, 1964, *Biochim. Biophys. Acta*, **90** 324
30. E. Kabat and M. Meyer, 1966, Experimental Immunochemistry, 2nd Ed., p. 264, Charles C. Thomas Publisher, Springfield
31. U.-B. Hansson and E. Nilsson, 1973, *J. Immunol. Met.*, **2** 221
32. J. Porath and B. Olin, *Biochemistry*, **22** 1621
33. C. Tanford, 1968, *Adv. Protein Chem.*, **23** 121
34. W. Jencks, 1969, Catalysis in Chemistry and Enzymology, p. 323, McGraw-Hill, New York
35. S. Timasheff and G. Fasman, 1969, Structure and Stability of Biological Macromolecules, Marcel Dekker, New York
36. F. Gurd and P. Wilcox, 1956, *Adv. Protein Chem.*, **11** 312
37. J. Hale and D. Beidler, 1994, *Anal. Biochem.*, **222** 29
38. J. Deisenhofer et al, 1978, *Hoppe-Seylar's Z. Physiol. Chem.*, **359** 975
39. E. Kabat et al, 1987, Sequences of Proteins of Immunological Interest, US Dept. of Health and Human Services, Public Health Service, National Insititute of Health
40. B. Compton et al, 1989, *Anal. Chem.*, **61** 1314
41. G. Blank and D. Vetterlein, 1990, *Anal. Biochem.*, **190** 217
42. S. Fulton et al, 1991, *BioTechniques*, **11**(2) 226
43. P. Grandics, 1994, *Am. Biotech. Lab.*, **Jun.** 12
44. A. Cripps et al, 1983, *J. Immunol. Met.*, **52** 197
45. M. Wickerhauser and Y. Hao, 1972, *Vox Sang.*, **23** 119
46. L.-G. Falksveden and G. Lundblad, 1980, in Methods of Plasma Protein Fractionation, (J. Curling, ed.), p. 93, Academic Press, London
47. Y. Hao et al,1980, in Methods of Plasma Protein Fractionation, (J. Curling, ed.), p. 57, Academic Press, London
48. T. Arakawa, 1985, *Anal. Biochem.*, **144** 267
49. J. Lee and L. Lee, 1979, *Biochemistry*, **18** 5518
50. J. Lee and L. Lee, 1981, *J. Biol. Chem.*, **156** 625
51. A. Ogston, 1958, *Trans. Faraday Soc.*, **54** 1754
52. P. Foster et al, 1973, *Biochim. Biophys. Acta*, **317** 505
53. A. Polson et al, 1964, *Biochim. Biophys. Acta*, **82** 463
54. I. Juckes, 1971, *Biochim. Biophys. Acta*, **229** 535
55. S. Miekka and K. Ingham, 1978, *Arch. Biochem. Biophys.*, **191** 525
56. S. Neoh et al, 1986, *J. Immunol. Met.*, **91** 231
57. K. Ingham, 1990, *Met. Enzymol.*, **182** 301

58. W. Honig and M.-R. Kula, 1976, *Anal. Biochem.*, **72** 502
59. — 1989, The Merck Index, 11th edition, Merck and Co., Rahway
60. P. Gagnon et al, 1995, A method for obtaining unique selectivities in ion exchange chromatography by addition of organic polymers to the mobile phase, poster, 15th International Symposium on HPLC of proteins, Peptides, and Polynucleotides, Boston, Reprint: <http://www. validated.com/library.html>, in press, J. Chromatogr., 1996, ref CHROMA 28044
61. D. Atha and K. Ingham, 1981, J. Biol. Chem., 256 12108
62. K. Ingham, 1978, *Arch. Biochem. Biophys.*, **186** 106
63. R. Almog and E. Schrier, 1978, *J. Phys. Chem.*, **82** 1701
64. P. Yang and J. Rupley, 1979, *Biochemistry*, **12** 2654
65. J. Steinhardt and M. Jones, 1978, in Physical Aspects of Protein Interaction, (N. Castimpoolas, ed.), p. 181, Elsevier, New York
66. A. Polson and C. Ruiz-Bravo, 1972, *Vox Sang.*, **23** 107
67. P. Gagnon, 1992, Multiple mechanisms for improving binding of IgG to protein A, poster, BioEast '92, Washington D.C., reprint: <http://www.validated.com/ library.html>
68. J. Brewer and L. Soderberg, 1977, in Chromatography of Synthetic and Biological Polymers, Vol. I, (R. Epton, ed.), p. 285, Ellis Horwood, Chichester
69. S.-C. Yan et al, 1984, *Anal. Biochem.*, **138** 137
70. K. Hellsing, 1968, *J. Chromatogr.*, **36** 170
71. J. Giddings and K. Dahlgren, 1970, *Sep. Sci.*, **5** 717
72. T. Laurent and J. Killander, 1964, *J. Chromatogr.*, **14** 317
73. L. Hagel, 1989, in Protein Purification: Principles, High resolution Methods, and Applications, (J.-C. Janson and L. Rydén, eds.), p. 63, VCH, New York
74. E. Fernandez et al, 1994, *Phys. Fluids*, **7**(3) 468
75. L. Plante et al, 1994, *Chem. Eng. Sci.*, **49**(14) 2229
76. T. Busby and K. Ingham, 1980, *Vox Sang.*, **39** 93
77. K. Ingham and T. Busby, 1980, *Chem. Eng. Commun.*, **7** 315
78. R. Scopes, 1982, Protein Purification, Springer-Verlag, New York
79. R. Steiner, 1953, *Arch. Biochem. Biophys.*, **46** 291
80. R. Steiner, 1953, *Arch. Biochem. Biophys.*, **47** 56
81. A. Plaut and T. Thomasi, 1970, *Proc. Nat. Acad. Sci. USA*, **75** 2649
82. M. Harboe and I. Fölling, 1974, *Scand. J. Immunol.*, **3** 471
83. A. Chantuin and R. Curnish, 1960, *Arch. Biochem. Biophys.*, **89** 218
84. M. Mckinney and A. Parkinson, 1987, *J. Immunol., Met.*, **96** 271
85. F. Perosa et al, 1990, J. *Immunol. Met.*, **128** 9
86. L. Reik et al, 1987, *J. Imunol. Met.*, **100** 123

87. M. Steinbuch and R. Audran, 1969, *Arch. Biochem. Biophys.*, **134** 279
88. J. McGregor et al, 1983, *Eur. J. Biochem.*, **131** 427
89. C. Russo et al, 1983, *J. Immunol. Met.*, **65** 269
90. H. Hoch and A. Chantuin, 1954, *Arch. Biochem. Biophys.*, **51** 271
91. E. Berry et al, 1956, *Arch. Biochem. Biophys.*, **62** 318
92. A. Habeeb and R. Francis, 1984, *Prep. Biochem.*, **14** 1
93. M. Temponi et al, 1989, *Hybridoma*, **8** 85
94. J. Ogden and K. Leung, 1988, *J. Immunol. Met.*, **111** 283
95. J. Lundblad and R. Seng, 1991, *Vox Sang.*, **60** 75
96. S. Timasheff and H. Inoue, 1968, *Biochemistry*, **7** 2501
97. R. Naito et al, US patent 4,446,134
98. L. Orö and A. Wretlund, 1991, *Acta Pharmacol. Toxicol.*, **18** 141
99. S. Gellis et al, 1948, *J. Clin. Invest.*, **27** 239
100. The calculations in reference 82 contain errors. ~0.2 mMoles of residual octanoic acid was associated with ~5 mMoles IgG after the first precipitation step.
101. D. Burton, 1985, *Mol. Immunol.*, **22** 161
102. G. Ballou et al, 1945, *J. Biol. Chem.*, **159** 11
103. Bio-Rad Laboratories, Hercules, CA USA
104. Supelco Inc., Bellafonte, PA USA
105. M. Stastny and J. Horejsi, 1961, *Clin. Chim. Acta*, **6** 782
106. A. Neurath and R. Brunner, 1969, *Experentia*, **25** 668
107. F. Franek, 1986, *Met. Enzymol.*, **121** 631
108. M. Steinbuch, 1980, in Methods of Plasma Protein Fractionation, (J. Curling, ed.) p.33, Academic Press, London
109. A. Horejsi and R. Smetana, 1956, *Acta Med. Scand.*, **155** 65
110. H. Schultze et al, 1965, *Immunochemistry*, **2** 273
111. E. Dietzel and H. Geiger, 1964, *Behringwerke-Mitt.*, **43** 129
112. A. Saifer and L. Lipkin, 1959, *Proc. Soc. Exptl. Biol. Med.*, **102** 220
113. D. Schroeder and M. Mozen, 1970, *Science*, **168** 1462
114. J. Horejsi and J. Korinek, 1972, *Zentralbl. Bakteriol. Parasitenkd. Infektionskr. Hyg., Abt.1: Orig., Reihe*, **220** 513
115. B. Lambrecht et al, 1991, *Vox Sang.*, **60** 207
116. J. O'Brien et al, 1990, *J. Lab. Clin., Med.*, **116** 439

Chapter 3

Size Exclusion Chromatography

"...it's a healthy thing now and then to hang a question mark on the things you have long taken for granted."
—Bertrand Russell

Size exclusion chromatography (SEC) of proteins was first described in 1955 and came into wide practice following the introduction of dextran-based media in 1959.[1,2] A variety of natural, synthetic, and hybrid polymer media have since been developed that embody a wide range of operating characteristics. SEC has been popular for lab scale purification of monoclonals—especially IgMs—since their emergence, but compared with other purification methods, it provides low resolution, low capacity, and it is slow. These limitations and their scale-up ramifications have restricted commercial use. Nevertheless its unique selectivity is sometimes required to achieve the necessary result.

Mechanism

A column bed has 3 functional components: the pore volume, the void volume, and the matrix volume (Figure 3.1). These components and their relationships define the separation capabilities and operating characteristics of a particular gel. Pore volume refers to the pore-lumen space within the particles. It is the sole determinant of selectivity and the primary determinant of capacity. Void volume is sometimes referred to as the interstitial volume or the excluded volume. It refers to the space between the particles. Void characteristics are the primary determinants of separation efficiency; how much a separation will be compromised by peak spreading. Matrix volume refers to the solid component of the particles. It is a codeterminant of mechani-

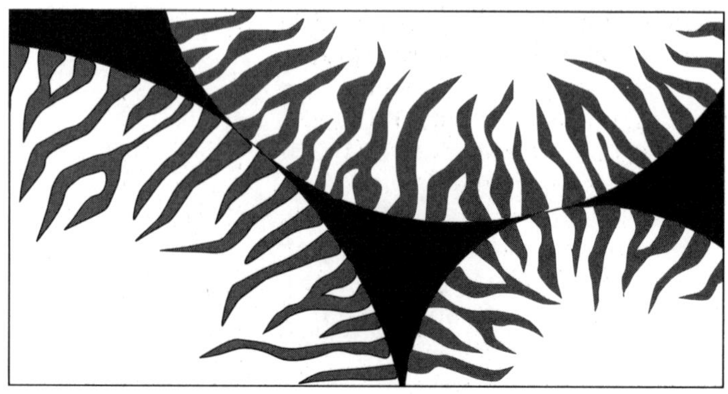

Figure 3.1. Functional components of a packed SEC column. The pore volume is indicated in gray, the void volume in black, and the matrix volume in white.

cal strength and porosity. The distinction between selectivity and separation efficiency, and their contributions to resolution, are illustrated in Figure 3.2.

the pore volume

The maximum pore diameter defines the exclusion limit of the gel. Proteins too large to enter the pores are said to be excluded. They have access to only the void volume. This provides them a shorter flow path to traverse than proteins that have access to both the void volume and the pore volume. Proteins larger than the exclusion limit elute together in a single peak at the beginning of the chromatogram: the void peak.

Proteins having access to the pore volume are said to be included. Their fractionation characteristics are determined by the pore size distribution within the pore volume.[3,4] Proteins at the upper end of the range are included only in the larger pores.This gives them access to a lesser fluid volume than smaller proteins, hence a shorter path length and earlier elution. Molecules small enough to be included in all the pores elute, undifferentiated from one another, at the end of the chromatogram.

The pore size distribution for a given gel can be represented by a calibration curve of molecules with known molecular size (Figure 3.3). Plotting the elution volume versus the molecular size of the standards yields a generally sigmoidal curve. The exclusion limit is indicated by the ascending arm, and the point of total inclusion by the descending arm. The useful fractionation range is indicated by the "linear" segment within the shaded area. The narrower the fractionation range,

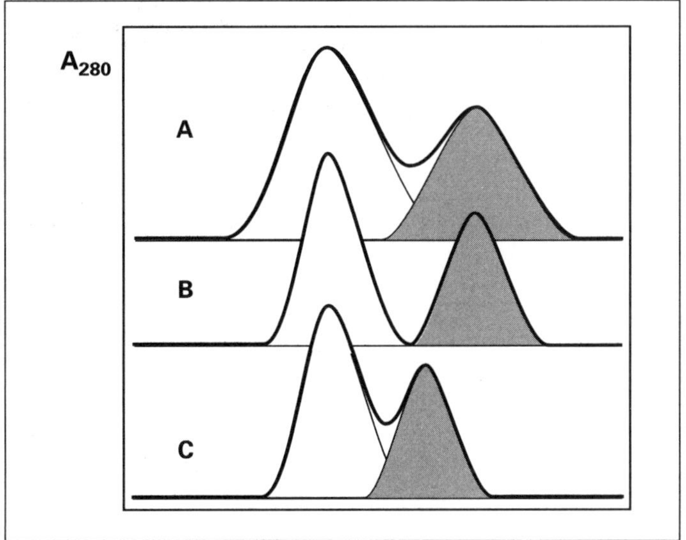

Figure 3.2. Selectivity, separation efficiency, and resolution. Profiles A and B represent columns with identical selectivity. Profiles B and C represent columns with identical separation efficiency. Profiles A and C represent columns that support identical resolution. Good selectivity and good separation efficiency are both required to achieve good resolution.

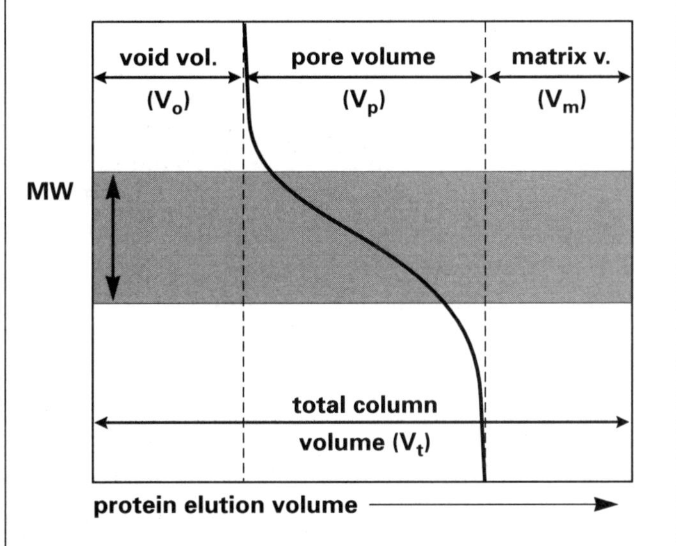

Figure 3.3. A size exclusion calibration curve. The shaded area indicates the useful fractionation range. The slope of the curve through this region indicates selectivity. The shallower the slope, the better the discrimination. Calibration curves offer no information concerning separation efficiency.

the larger the proportion of pores of a given size within the range. Among gels of equivalent pore volume, this translates into finer selectivity, as well as higher capacity.[3,4] In general, the proportions of various pore sizes within the gel follow a Gaussian distribution, with the largest proportion residing at the inflection point.[5] Selectivity and capacity are therefore best near the middle of the range.

conformational effects

The log of separation volume is proportional to the log of molecular size.[6-9] Linearity versus molecular weight is good within a conformational class of molecules, for example among linear polymers or among globular proteins, but poor across conformational classes. Aspheric molecules behave as spheres with diameters equal to their longest dimension. The hydrodynamic volume of a rod shaped molecule can be triple that of a sphere of equal mass.[10] This is important for antibodies, IgMs being discoid and other classes being pseudo-linear. This causes up to a 40% overestimate of IgG molecular weight.[11]

efficiency of mass transport

Separation efficiency is strongly dependent on flow rate.[3,4] The molecules to be fractionated must have an opportunity to diffuse into and out of the pores, in equilibrium with the transport rate of the sample down the column. Proteins unable to penetrate the pores cause broadening to the leading side of the peak. Proteins that exit the pores after passage of the main sample body cause broadening to the trailing side. These effects are described as reflecting poor mass transport efficiency. Mass transport efficiency improves with reducing flow rate—to a point. Running a column too slowly allows peak spreading from simple diffusion. This is called axial dispersion and can cause more loss of performance than is gained from improved mass transport efficiency.[3,4] The best separation performance is achieved when the combined dispersion from the 2 phenomena is at minimum.

restricted diffusion

Rates of protein diffusion inside gel pores are reduced by restrictions on their range of motion, but remain proportional to mass.[12-14] Table 3.1 illustrates diffusivities calculated for antibodies with an equation validated to support good agreement with experimental SEC data for other proteins.[12,13,15] These values translate proportionally to optimal flow rate.[12,16,17] The best flow rate for IgM is generally about half the rate for IgG.

the void volume

The quality of separation achieved by the pore volume is eroded by the void volume. Void flow rate is reduced by friction near particle surfaces. This causes

Table 3.1. The relationship between protein diffusion constant and nominal flow rate for gels with different particle diameters. K_{dif} indicates diffusion contant in cm^2/sec. Flow rates are given in cm/hr. α2M indicates α_2-macroglobulin. The equation for calculating diffusion constants was taken from reference 15. The equation for nominal flow rate was taken from references 3 and 5.

protein	mass	K_{dif}	diam µm/flow 10	30	100
light chain	23kD	$9.1x10^{-7}$	35	12	3.5
albumin	67kD	$6.7x10^{-7}$	25	8.5	2.5
IgG	150kD	$4.9x10^{-7}$	19	6.0	2.0
IgA	160kD	$4.8x10^{-7}$	19	6.0	2.0
IgA	335kD	$3.7x10^{-7}$	14	5.0	1.5
IgD	170kD	$4.7x10^{-7}$	18	6.0	2.0
IgE	190kD	$4.5x10^{-7}$	17	5.0	2.0
α2M	725kD	$2.9x10^{-7}$	11	3.5	1.0
IgM	960kD	$2.6x10^{-7}$	10	3.0	1.0

turbulent mixing in the interstices, referred to as eddy dispersion. The effect is proportional to the void volume, increasing linearly with particle diameter. It degrades overall mass transport efficiency by the square of particle diameter. Among media of equivalent mean diameter, those with broader particle size distributions have larger void volumes and cause more eddy dispersion. Eddy dispersion is independent of flow rate.

packing aberrations

Eddy dispersion is amplified by packing aberrations. Particles occur in lower density at column walls. This creates a boundary layer of about 3 particle diameters in which the flow velocity is substantially higher than at the column center.[18] The discontinuity creates an elevated dispersion layer that extends inward about 30 particle diameters from the wall. The compromise to performance with large particles in narrow columns can be severe. In a 1 cm diameter column packed with 60µm particles, the wall dispersion zone constitutes 59% of the bed. This zone is reduced to 26% in a 2.6 cm column, and 14% in a 5 cm column. These percentages vary in direct proportion to particle diameter.

collective dispersion

Figure 3.4 illustrates an example of the collective effects of dispersion on separation efficiency, based on the classical van Deemter equation.[19] Although developed originally with gas chromatography, it has

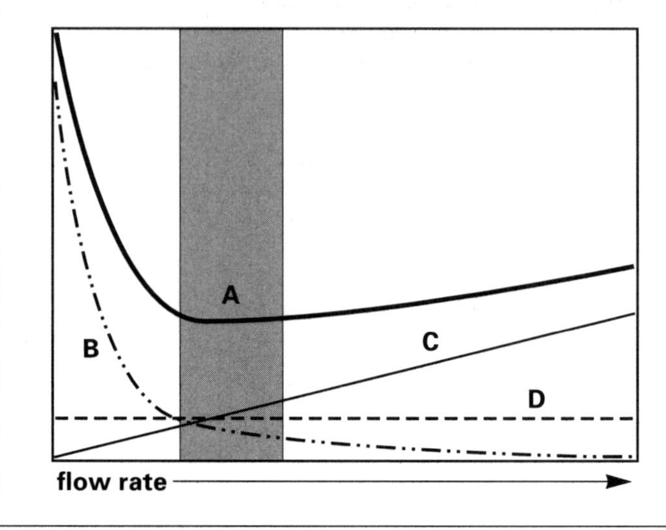

Figure 3.4. Cumulative effects of dispersion on separation efficiency. Curve A integrates the effects of the following sources of dispersion. Curve B illustrates axial dispersion. Curve C illustrates loss of mass transport efficiency as a function of flow rate. Curve D illustrates eddy dispersion. The shaded area indicates the range of flow rates supporting optimum performance. Curves like this do not provide any information concerning selectivity.

been verified to accurately represent liquid chromatography.[16,20]. The same protein will produce different plots on different gels according to differences in their pore and void volume characteristics. Different proteins will produce different plots on the same gel according to their diffusivities, which will also vary with mobile phase formulation and temperature.

matrix volume

Matrix volume is a codeterminant of a gel's resistance to flow, its mechanical strength, and its capacity. The lowest resistance to flow and highest capacity come with the highest ratio of pore volume to matrix volume. This requires a rigid base polymer. If the base material is inherently weak then mechanical strength can be improved only by increasing matrix volume at the expense of the pore volume. This sacrifices capacity and increases resistance to flow.

operating formats

SEC is employed in 2 formats: high resolution fractionation, which is used for separation of proteins from one another; and buffer exchange chromatography, which is used to perform gross separations of protein from salts and other small molecules. This chapter will focus on high resolution applications, although most of the points apply equally to buffer exchange.

Attributes

The most compelling feature of SEC is its simplicity. Since performance is determined predominantly by

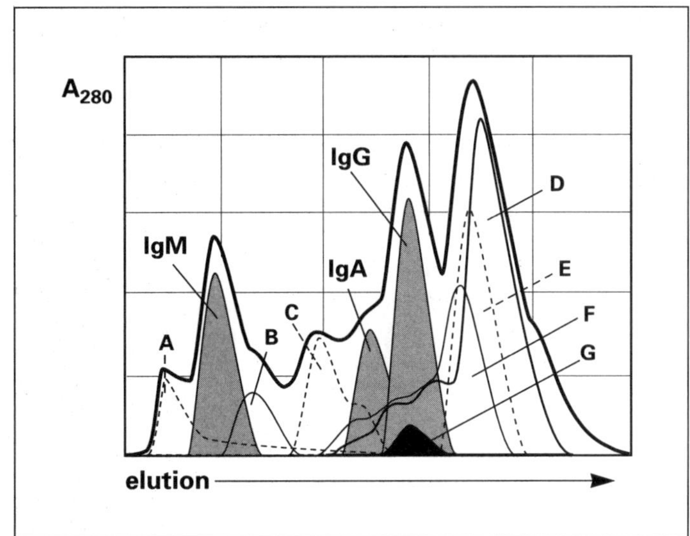

Figure 3.5. Relative elution positions of antibodies and major protein contaminants.
A: lipoprotein
B: α_2-macroglobulin
C: haptoglobin
D: albumin
E: transferrin
F: complements
G: β_1-haptoglobin
Consult references 23 and 24 for more information on contaminant proteins.

the gel, method development is minimal. Equipment requirements are likewise simple, with no requirement for sophisticated gradient capability. At its most primitive level, SEC can be conducted in open columns under gravity flow, but this does not provide adequate performance for most commercial applications.

purification performance

SEC provides reasonable-to-good purity from raw feedstreams; 50-80% for IgGs, and 80–90% for IgMs. Baseline resolution requires approximately a 2-fold size differential between proteins.[3,4] This makes SEC an effective tool for removal of antibody aggregates but it provides no discrimination of the target monoclonal from nonspecific antibodies of the same class.[21-25]

typical contaminants

Other proteins commonly contaminating IgG preparations include C3, C4, haptoglobin, β_1-haptoglobin, ceruloplasmin and inter-α-trypsin inhibitor.[22,23,26] Typical IgM contaminants include α_2-macroglobulin, HDL and LDL (Figure 3.5).[26-28] Some SEC media have sufficiently broad fractionation ranges to accommodate both IgG and IgM, but better purity and capacity are obtained from media with narrower ranges matched to the size class of the particular antibody).

DNA removal

The ability of SEC for DNA removal is dependent on the size distribution of the DNA. A fraction of the DNA inevitably coelutes with the product but overall

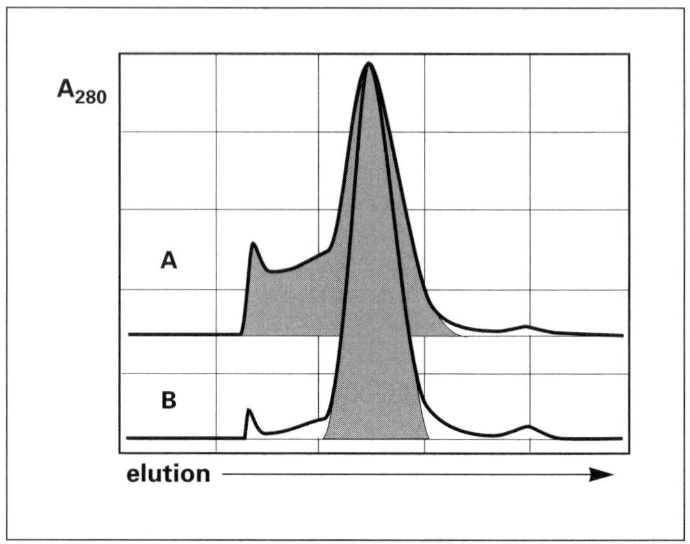

Figure 3.6. SEC of IgM complexed with DNA (A) and after decomplexation (B). Profile A was run in 0.05M sodium phosphate, 0.10 sodium chloride, pH 7.0. Profile B shows the same sample pre-equilibrated and loaded with 1.0M sodium chloride. Shaded areas indicate the elution position of the antibody. Besides carrying DNA through a process, complexation depresses recovery. The "monomeric" peak from the complexed profile contained less than 70% of the antibody. Many antibodies exhibit this type of behavior but usually not to this extreme.

reductions of 1–2 logs are common. An important exception results from the ability of DNA to form stable ionic complexes with antibodies at low conductivities. Under these conditions, the DNA can be carried through SEC with negligible reduction (Figure 3.6).[29] These complexes can be dissociated and the DNA largely removed by increasing the ionic strength of the buffer. IgMs and strongly basic IgGs may require up to 1.0M sodium chloride for dissociation, but 0.3M is sufficient for most antibodies.[29-31]

Another important caution with DNA: highly contaminated process streams are viscous. Samples with viscosity more than 50% higher than the mobile phase are hydrodynamically unstable.[3] The mobile phase tends to punch holes through the sample layer, creating multiple subzones. This phenomenon is called fingering because proteins that normally elute in single peaks elute in broad multi-peak zones, like the fingers of a hand.[3,32,33] Fingering can be eliminated by increasing the mobile phase viscosity to match the sample but this requires reduction of flow rate, not only to control backpressure but to compensate for viscosity-mediated reduction of protein diffusivities.[3,34]

virus removal

SEC supports virus removal of 1.5–3.5 logs.[35-37] These values are modest compared with other purification methods, but it's important to remember that the selectivity is unique and may provide a necessary overall boost to a process. As with other methods, it's important to rigorously sanitize the gel after each use to prevent carry-over of residual virus to subsequent product lots.[37]

endotoxin removal

SEC supports 1–2 logs of endotoxin removal for IgG; less for IgM. Under physiological salt and pH conditions, endotoxins exist in secondary structures of 300–1000kD.[38] Larger structures are formed with increasing concentrations of lyotropic salts, shifting endotoxin populations toward the void peak.

mass recovery

Mass recovery is typically at least 90%. If less, it usually indicates that buffer formulation needs to be modified to eliminate one or more of 3 problem sources: product insolubility, product complexation with other sample components, or product interactions with the gel media.

activity recovery

Activity recovery is typically quantitative with SEC, owing to the inertness of the gel and the gentleness of the separation conditions. This is in contrast to methods that require product exposure to extreme pH, conductivity, or strongly adsorptive ligands.

buffer exchange capability

The ability of SEC to exchange buffer formulation coincident with fractionation makes it a useful process design tool; for example, equilibrating a high-salt HIC pool to conditions more appropriate for ion exchange (IEC). Care must be taken to ensure that the antibody is fully soluble under the destination conditions. It's also important to be aware that high concentrations of precipitating salts or organic additives such as polyethylene glycol may alter SEC separation performance.[39,40] This is usually not an issue with resuspended precipitates but it can be a problem with supernatants.

resistance to harsh sanitizing conditions

Polymeric SEC media are quite robust. Most will withstand acids, bases, chaotropes, detergents and organic solvents. With careful sample preparation, regular cleaning, and occasional frit replacement, analytical SEC columns endure thousands of injections without loss of performance.[41] Preparative columns have shorter lifespans, mainly as a result of fouling from increased sample exposure. Even so, a given batch of media should last 50–100 applications.[42]

Limitations

The reliance of SEC on diffusion as a separation mechanism makes it vulnerable to secondary interactions that alter protein-diffusive behavior. This would be a minor concern if SEC media were as inert as most users assume. Unfortunately, most gels fail this standard. SEC media have been reported to participate in charge interactions, hydrophobic interactions, hydrogen bonding, and biospecific interactions.[7,28,43-59]

charge interactions

Charge interactions are a concern because they are influential within the range of buffer conditions typically used for SEC. Agarose-based media often bear residual sulfates, while dextrans carry residual carboxyls, and polyacrylamides carry primary amino groups.[3,60,61] Acid or alkaline hydrolysis due to extended exposure to pH extremes can also introduce charged groups into gel media.[3] Some media bear up to 0.2 milliequivalents of charge groups per mL of gel.[62] This, and much lower charge densities, can significantly influence selectivity and separation performance. Proteins with the same charge as the matrix are partially excluded from the pores and elute earlier than expected in sharper peaks. Proteins with the opposite charge are either retarded or retained and elute later in broader peaks.[3,4] Charge interactions can be suppressed with 0.2–0.5M sodium chloride.[3,48,55]

hydrophobic interactions

Hydrophobic sites on SEC media may derive from crosslinking agents used in polymerization, surface coatings, or the inherent hydrophobicity of the base matrix.[3] Interactions with these sites retard elution, causing antibodies to elute later than expected, usually in wider peaks. Hydrophobic retardation causes the molecular weight of IgG to be underestimated by 30–35% on some media.[52] This is remarkable considering that the linear conformation of antibodies causes roughly equivalent overestimates on noninteractive media (Figure 3.7).[11]

hydrogen bonding

SEC media are rich in hydroxyls—potential hydrogen donors—and hydrogen bonding is believed to be a significant contributor to nonspecific binding.[39,50,

Figure 3.7 Elution volume of 2 monoclonal antibodies as a function of ammonium sulfate concentration. Elution volume (V_e) expressed as % of total column volume. Phosphate has a similar effect. Nonprecipitating salts like sodium chloride have relatively little effect.

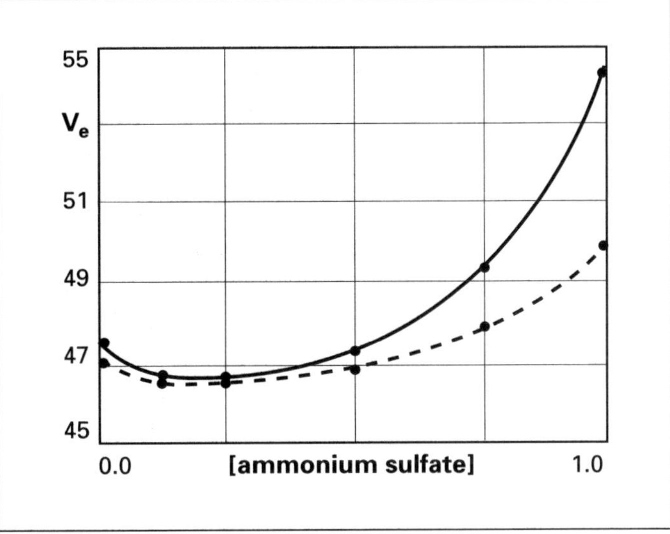

58,59] Hydrogen bonding can be blocked by the inclusion of hydrogen donor/acceptors such as urea or sucrose in the buffer formulation.[59,63-65] Concentrated sucrose is viscous and may require reduced flow rates, so urea is generally the better choice. Denaturation concerns are negligible at the required concentration: ~1.0M.[64-66]

antibody solubility limitations

Besides blocking electrostatic gel:protein interactions, moderate salt concentrations are required to maintain antibody solubility. Most IgMs and many IgGs are poorly soluble at low ionic strength, especially at low pH. IgMs have been reported to adsorb to SEC media under these conditions (28,67,68). Borderline salt concentrations result in broad poorly defined elution peaks. Like salt, glycine has a solubilizing effect on antibodies, especially IgMs.[69-71] One of glycine's advantages is, that being zwitterionic, it doesn't interfere with downstream charge-based separations. However, it does strip nickel from immobilized metal affinity (IMAC) media.[72]

mobile phase enhancement of nonspecific interactions

Excessive concentrations of precipitating salts cause adsorption of hydrophobic proteins.[58,59] Even moderate concentrations may cause retardation and peak broadening (figure 3.7).This causes antibodies to coelute with smaller contaminants.

Figure 3.8. The effect of packing flow rate on column performance. Resolution was measured between albumin and myoglobin on 2 semi-rigid SEC supports. Data replotted from reference 73.

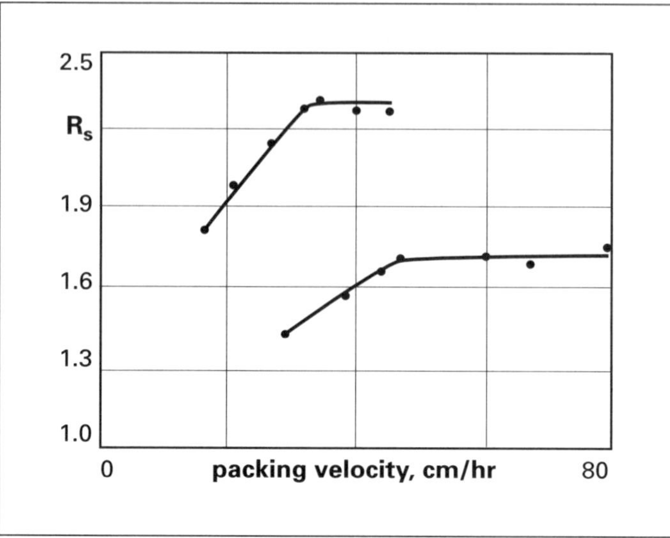

dependence on packing quality

The dependence of SEC on well-packed beds makes it more vulnerable than other methods to variations in packing technique. This is not to suggest that other methods don't require a well-packed bed. They do, but their separation characteristics derive predominantly from controlled changes in mobile phase composition. In addition, the long beds required for SEC are more difficult to pack well than the short columns required for adsorption chromatography.

A thorough treatment of column packing is beyond the scope of this chapter. It's best that you refer to manufacturer's instructions. However, a few points deserve particular attention. Packing quality is often improved at higher packing flow rates so long as the pressure doesn't compress the particles (Figure 3.8).[3,73] Since pressure drop is proportional to bed height, you can maximize packing flow rates by packing shorter column segments and subsequently plumbing them together. This can be advantageous even for small diameter columns, but becomes especially useful with large diameters. Larger diameters provide less wall support to the bed, reducing its pressure resistance.[74] It also becomes increasingly costly to maintain precision diameter and adequate wall strength for long column tubes with large diameters.

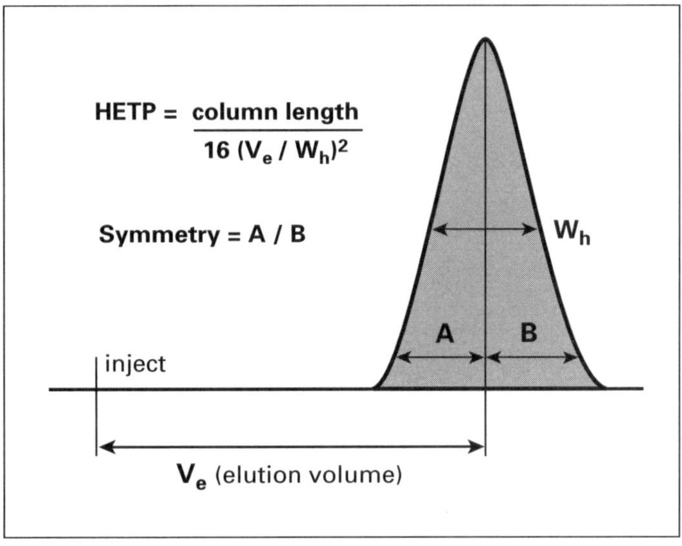

Figure 3.9 Evaluation of packing quality by HETP and symmetry factor. W_h refers to peak width at half-height. A and B are measured at 10% of the peak height.

evaluation and documentation of packing quality

The sensitivity of SEC to packing variations makes documentation of packing quality essential in a regulated manufacturing environment. Most process guides and book chapters recommend using the height equivalent of a theoretical plate (HETP) as an index (Figure 3.9). The HETP is determined for an inert low molecular weight tracking substance by measuring the elution volume and width of the eluted peak, then relating them back to the column volume. An exceptionally well packed column will yield an HETP of about 1.5 times the mean particle diameter.[12,13] Values twice the average diameter are considered good.[75] Other variables being equal, better HETPs are obtained from media with narrower particle size distributions.

Peak symmetry provides a complementary expression of packing quality. It can be calculated from the same chromatogram used to calculate HETP. Symmetry provides a quantitative expression of peak skew; leading or tailing. Pronounced asymmetry may indicate a problem with flow distribution at the inlet due to fouling, or poor flow distribution through the bed due to channeling or local bed compression.[3] Symmetry factors for small molecules on a well packed column should be between 0.9 and 1.1.[3]

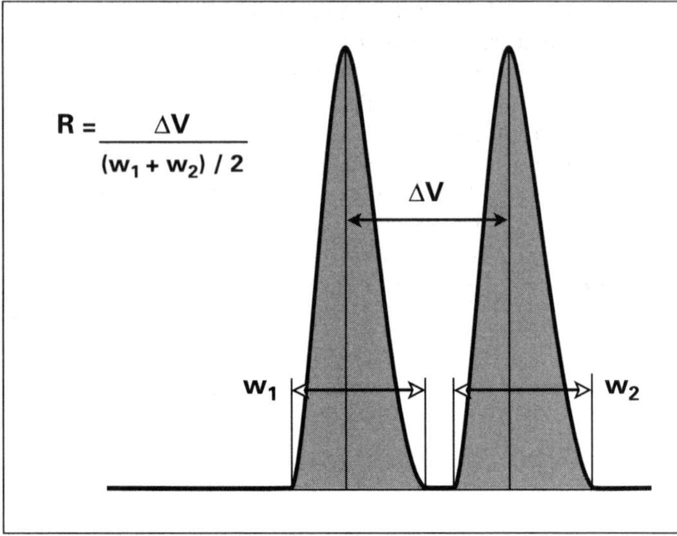

Figure 3.10. Calculation of resolution between adjacent peaks. Baseline resolution is generally indicated by a resolution factor of ~1.5.

limitations of standard measures

A weakness of HETP and symmetry measures is that they don't reflect secondary interactions that may alter separation performance. A more meaningful way to measure packing quality is to use a sample of your antibody combined with a protein of about half its size and another about double. Load a specified volume of sample at a specified linear flow rate and evaluate the separation versus a control run on a column known to provide adequate separation performance.

Besides incorporating the influence of secondary interactions, this approach allows you to directly measure variations in resolution (Figure 3.10). You should still calculate HETPs and symmetry factors, but evaluate them against the control, not against standards produced with small molecules. HETPs vary with the diffusivity of the molecule, so plate heights for proteins are much larger than for small molecules. Likewise, symmetry factors for proteins rarely match the ideal of an inert low molecular weight tracking substance.

ex-column performance factors

The sensitivity of SEC performance to dispersion extends beyond the column to include the chromatography system. The most critical function is sample application.[3] The best systems either employ or mimic the pressurized injection loop system characteristic of HPLC. Appropriate valves can be obtained for

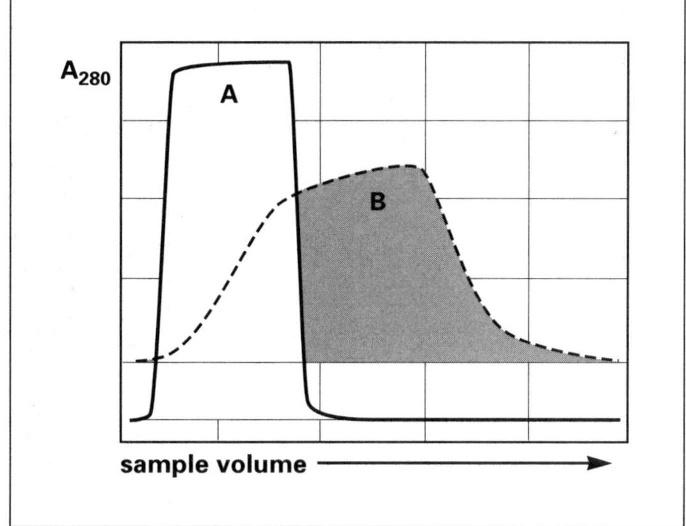

Figure 3.11. Precolumn sample dispersion as a function of sample application. Profile A was generated by loading sample directly to a UV monitor through a pressurized injection loop. B was loaded through a peristaltic pump. Initial sample volume and concentration were the same for both traces.

any production scale. They are relatively expensive but worth the investment. It is also necessary to optimize configuration of the sample loop itself. Loop diameter should be uniform throughout, no greater than the diameter of the tubing leading into it, and as short as possible. Preparative sample volumes (0.5–150 mL) can be injected from a Superloop.[3] Functional analogs can be fabricated for very large sample application.

sample application systems

Three sample application systems to avoid are very long tubing loops, "balloon loops", and through-the-pump sample application. Long sample loops dilute the sample by laminar dispersion. Balloon loops are sample loops with narrow-bore tubing at the valve connections but a larger diameter through the main body to increase volumetric capacity. The discontinuity in diameters makes these devices extremely effective turbulent mixing chambers, diluting the sample boundaries, increasing overall sample volume, and devastating separation performance. Applying sample through a chromatograph's primary pumping system has the same effect. Whatever sample application system is employed, it should be placed as close to the inlet of the column as possible.[3] The worst thing you can do is place a bubble trap between the point of sample introduction and the column.

Figure 3.12. Resolution as a function of sample volume. Sample: albumin and mouse IgG_1, each at 2.5mg/mL. Albumin was prefractionated to remove polymers. Flow rate: 20cm/hr.

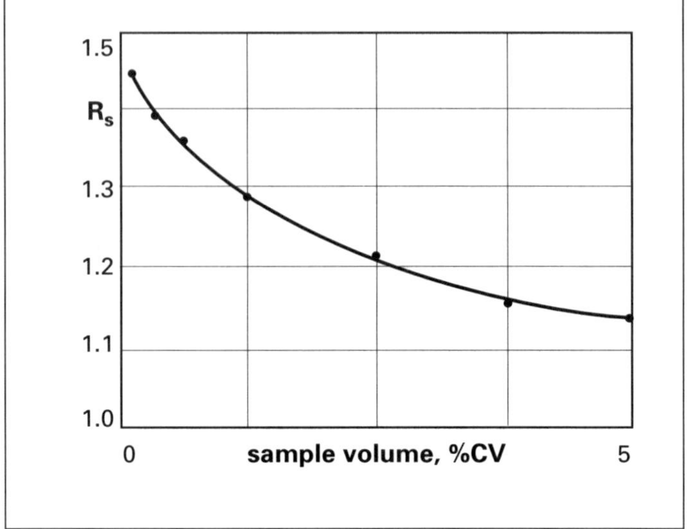

evaluating sample application systems

You can evaluate the relative abilities of different sample application systems by injecting a mock sample of 1% acetone through the chromatograph with the column off line (Figure 3.11). If the system fails to maintain near-vertical sample boundaries, that's a warning sign that it will detract from column performance. Make sure the sample application system supports equivalent performance at all process scales in advance of scale-up.

other sources of dispersion

Equal care must be taken to control dispersion between the column and the fraction collector. Minimize the number of tubing segments, make sure that they are of uniform internal diameter, and use zero dead-volume fittings at all junctures. One of the most insidious detractors from performance is poor flow cell design in the detector. Even in some of the newer detectors, the flow cell balloons out in the middle. This causes turbulent mixing, robbing the separation accomplished by the column. Replace such cells with uniform diameter flow cells. Replace detectors that won't accommodate such cells.

low capacity

SEC has the lowest capacity of the major fractionation techniques.[4] HPLC media (5–10 µm) will tolerate injections of only 0.3–0.5% of the column volume (CV) without loss of performance (Figure 3.12).[3,76]

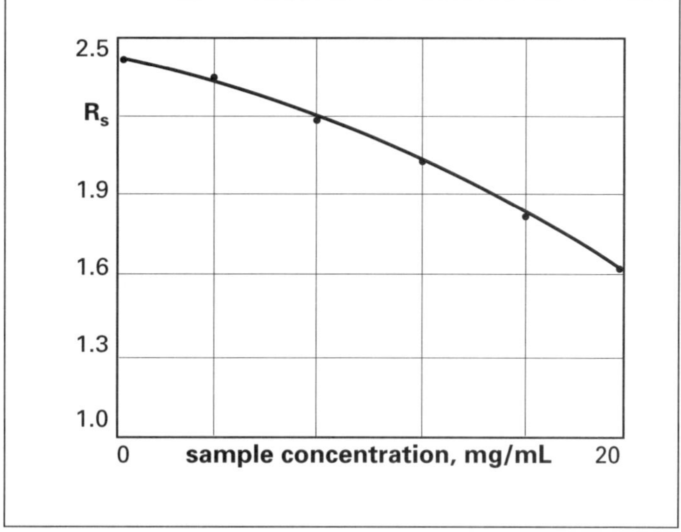

Figure 3.13. Resolution as a function of sample concentration. Sample: albumin and mouse IgG_1, in equal mass ratios. Albumin was prefractionated to remove polymers. 1% of CV was loaded at 20cm/hr.

Preparative media may support injections up to 4% CV.[3] Contrast this with adsorption techniques that support sample applications many times the column volume. This translates into larger volumes of SEC media to achieve comparable productivity, and a disproportionate increase in material expense.

effect of sample concentration on resolution

SEC capacity can be improved somewhat by concentrating the sample, but the elevated viscosity produces a steady decline in resolution (Figure 3.12).[71] If sample concentration requires a separate process step, you also need to factor in the increased processing costs and inevitable product losses.

long separation times

SEC is the slowest of the fractionation techniques. High performance analytical scale applications can be conducted in 30 minutes or less, but even this compares poorly with adsorptive methods that can be performed with the latest generation media in less than 5. Preparative applications can be agonizingly slow; seldom less than 4 hours, commonly 8–12, sometimes more .[77] This is expensive. Low throughput ties up resources and requires additional media, equipment systems, and more manufacturing space to achieve throughput equal to other methods. If these resources are unavailable, then reliance on SEC imposes an artificial limit on production capacity.

temperature considerations

Because of SEC's slowness and concerns about product loss or modification due to enzyme activity, many users conduct their separations in the cold. Keep in mind that a reduction in temperature from 22°C to 4°C increases viscosity of aqueous solutions by a factor of about 1.5.[3] This approximately doubles the pressure drop and reduces diffusivity, requiring reduction of flow rate to maintain performance.[3] If you plan to run a process in the cold, develop it in the cold.

Temperature is a special concern with antibodies because some of them are cryoglobulins; progressively insoluble below 37°C. This is an occasional problem with IgGs, but it affects up to 20% of IgMs.[63,78-80] With some, it can be a factor even at room temperature. Others may become problematic only at cold temperatures. Regardless of the threshold, temperature adopts obvious importance as a process control variable.

Cryoprecipitation is driven mainly by polar interactions, both salt-bridging and hydrogen bonding.[63,78,79] It is relatively independent of pH from 5–10.[80] Increased conductivity will suppress it in some cases, but at such high concentrations that enhancement of hydrophobic interactions becomes a concern. Strong hydrogen donors such as urea and sucrose are the most effective blockers.[64-66] As noted earlier, 1.0 M urea is the more appropriate candidate because of its lower viscosity.

A final warning: cryoprecipitates are often transparent and nearly invisible on container surfaces. This can lead to inadvertent product losses in a wide variety of circumstances. With antibodies you identify as cryoglobulins, make sure that SOPS specify that samples must be raised to an appropriate temperature prior to transfer between vessels, and especially before conducting any separation process.

Method development

The reliance of SEC performance on gel characteristics makes media selection the core component of method development. Unfortunately, commercial SEC media are not indexed in way that readily supports identification of the best candidates for your needs. Even if a valid indexing scheme was possible regarding the physical characteristics of various gels, it would still be impossible to predict the effects of solute:gel interactions, solute:solute interactions, or their respective responses to variations in the mobile phase. The only way to qualify gel media is to evaluate them directly.

media evaluation

A sample composed of your own monoclonal with proteins about half and about twice its mass is well suited for preliminary screening. Use dilute samples of ~0.5% of the column volume to conserve sample. The running buffer should be formulated to block known secondary interaction mechanisms. Begin with neutral pH, and 0.5M sodium chloride. This should suppress ionic interactions without enhancing hydrophobic interactions, and avoid antibody solubility limitations. Include 1.0M urea to suppress hydrogen bonding. 1.0M glycine may be helpful with IgMs.

capacity comparison

Once you have identified gels that provide appropriate selectivity, you need to compare their relative capacities. Prepare columns of equivalent bed height and diameter, then measure resolution for each at sample loads of 4% CV. Measure resolution to discriminate performance. Higher resolution translates into higher capacity, so pick the gel that gives the best resolution.

process order

Size exclusion is not well suited as a first process step, primarily because the antibody is a minor component of the total sample mass. The technique's reliance on homogeneity of flow through the bed also makes it especially vulnerable to foulants that increase the probability of channeling. Once a preliminary process sequence has been established, compare separation performance in the destination buffer formulation against performance in the blocking formulation recommended above. If nonspecific interactions in the destination buffer detract excessively from separation performance, then it must be reformulated or SEC abandoned. Even if the interactions appear moderate, you risk uncontrolled process variability from differences among gel media lots.

bed height

There is a natural tendency to increase bed height to augment separation performance, but it's important to keep in mind that separation improves only by the

pore volume

square root of the increase while product dilution increases in direct proportion.[3] The positive aspect of this relationship is that minor variations in bed height have negligible effect on performance and reproducibility. Most preparative high resolution applications use bed heights of 60–90cm. Most buffer exchange applications employ bed heights of 5-15cm.

setting capacity specifications

After the final process sequence has been set and material is available that accurately reflects sample composition for the intended final process, then capacity specifications can be set. This primarily involves varying the relative volume of the sample load (Figure 3.13). If the preceding step yields highly concentrated material then you may benefit from experimenting with sample concentration as well.[3,76] When setting specifications, be sure to accommodate variations in packing quality that may occur in Manufacturing.

scale-up issues

The majority of scale-up problems with SEC derive from poor column packing. Give packing methods the attention they deserve as a major determinant of process performance. Consult with media manufacturers, experiment with different methods, and do whatever it takes to make sure that you are able to consistently achieve the packing quality you require at all process scales. As noted previously, consider linking short column segments, even for columns with small diameters. The improvement in packing quality and the relative ease of achieving it are significant.

maintenance

Achieving good packing quality is one thing, maintaining it is another. A frequent source of manufacturing process failures is dehydration of the gel bed. Even if a column is well sealed by laboratory standards, the overwhelming molecular forces of evaporation eventually pull water from the bed. This causes channeling. In severe cases you can see cracks, but at the early stages the gel may only appear to have pulled away from the column walls at the top of the column. Adding buffer may restore the appearance of the gel, but its performance is altered permanently. This would be bad enough if it only caused the column to be repacked, but it often goes unrecognized.

This problem is worst with gels packed at room temperature for cold room use. When moved from room temperature into the cold, the fluid and seals contract. This exerts tension and air is pulled around the seals. Even if the seals hold, the gel bed will still shrink away from the walls due to fluid contraction. One solution is to pack the columns in the cold but this still leaves them vulnerable to eventual dehydration. A solution that addresses both issues is to connect a pressurized fluid reservoir to the column inlet. This need be nothing more than a spring-loaded oversized syringe. Make sure that there is a maintenance SOP calling for periodic inspection of reservoir fluid levels.

The other major source of scale-up problems is variation in the performance of the sample application system. As discussed earlier, this should be evaluated, refined as necessary, and validated before scale-up.

In perspective

SEC's unique selectivity makes it one of the foundation tools for monoclonal purification. Its role is supported by its ability to link otherwise incompatible process steps such as HIC and IEC. On the other hand its slowness, low capacity, and sample dilution effects are serious burdens. Even its abilities as a processing intermediate are compromised by antibody solubility limitations and secondary interactions between proteins and the gel.

purification of in vitro products

SEC is at its best with IgM purification. For diagnostic products, 2-step combinations with SEC following HIC, IEC, salt precipitation, or PEG precipitation all give good results. Such combinations are sometimes adequate for purification of IgGs, but not competitive with 2-step adsorptive chromatography alternatives. Most of the time, high resolution SEC does not contribute sufficient added purification value to justify its inclusion as an intermediate buffer equilibration step—at least not for in vitro products. If intermediate equilibration is required, simple buffer exchange chromatography is usually the better option.

purification of in vivo products

SEC is more likely to be a valuable component in purification of antibodies for in vivo applications. This is due to its viral clearance properties and ability to link process steps. However, IMAC is challenging it as a faster, higher capacity alternative.

Recommended reading

The chapter by Hagel is one of the most thorough and best integrated general references on the subject of SEC.[3] For more on packing, refer to references 3, 73, 74, 81-83 and specific manufacturers' instructions. The article by Pfankoch provides a general procedure for evaluating nonspecific gel effects.[48] Factors affecting development of sample application conditions are discussed in detail in reference 76. Applications of SEC to purification of monoclonal antibodies are discussed in references 24, 27, and 84-90.

References

1. G. Lathe and C. Ruthven, 1955, *J. Biochem.*, **60** xxxiv
2. J. Porath and P. Flodin 1959, *Nature*, **183** 1657
3. L. Hagel, 1989, in Protein Purification; Principles, High Resolution Methods, and Applications, (J.-C. Janson and L. Rydén, eds.), p.63, VCH Publishers, New York
4. F. Regnier, 1983, *Science*, **222** 245
5. L. Hagel, 1988, in Aqueous Size Exclusion Chromato-graphy, (P. Dubin, ed.), p.119, Elsevier, Amsterdam
6. H. Determann 1969, in Advances in Chromatography, (J. Giddings and R. Keller, eds.) Vol. 8, p.35, Marcel Dekker, New York
7. P. Andrews, 1970, in Methods of Biochemical Analysis, (D. Glick, ed.) Vol. 18, p.2, Wiley Interscience, New York
8. E. Cassassa, 1967, *J. Poly. Sci. Polym. Lett.*, **5** 773
9. A. Basedow et al, 1980, *J. Chromatogr.*, **192** 259
10. W. Yau and D. Bly, 1980, in Size Exclusion Chromatography (GPC), (T. Provder, ed.) p. 197, ACS Symposium Series, American Chemical Society, Washington D.C.
11. P. Andrews, 1964, *J. Biochem.*, **96** 595
12. J. Knox and F. McLennan, 1979, *J. Chromatogr.*, **185** 289
13. J. Leypoldt et al, 1984, *J. Appl. Polym. Sci.*, **29** 333
14. M. van Kreveld and N. vand den Hoed, 1978, *J. Chromatogr.*, **149** 71
15. C. Tanford, 1961, Physical Chemistry of Macromolecules, Chapter 6, Wiley, New York
16. J. Giddings and K. Mallik, 1966, *Anal. Chem.*, **38** 997
17. C. Tanford 1961, Physical Chemistry of Macromolecules, Chap. 6, Wiley, New York
18. J. Knox et al, 1976, *J. Chromatogr.*, **122** 129
19. J. van Deemter et al, 1956, *Chem. Eng. Sci.*, **5** 271
20. E. Katz et al, 1983, *J. Chromatogr.*, **270** 51
21. H. Suomela et al, 1984, *J. Chromatogr.*, **297** 369
22. F. Putnam, 1984, in The Plasma Proteins, (F. Putnam,ed.), Vol. 4. Chap.1, Academic Press, New York

23. H. Schultze and J. Heremans, 1966, Molecular Biology of Human Proteins, Vol.1, Elsevier, New York
24. S. Ikeyama et al, 1986, *Molec. Immunol.*, **23** 159
25. G. Sann et al, 1983, *J. Immunol. Met.*, **59** 121
26. J. Fahey and E. Terry, 1979, in Handbook of Experimental Immunology, (D. Weir, ed.), Vol.1, p. 8.1, Alden Press, Oxford
27. A. Jehanli and D. Hough, 1981, *J. Immunol. Met.*, **44** 199
28. J.-P. Bouvet et al, 1984, *J. Immunol. Met.*, **16** 299
29. P. Gagnon, 1996, Special Weapons and Tactics for Removal of Product-bound DNA, slide presentation, BioEast '96, Washington D.C.
30. R. Steiner, 1953, *Arch. Biochem. Biophys.*, **46** 291
31. R. Steiner, 1953, *Arch. Biochem. Biophys.*, **47** 56
32. E. Fernandez et al, 1994, *Phys. Fluids.*, **7**(3) 468
33. L. Plante et al, 1994, *Chem. Eng., Sci.*, **49**(14) 2229
34. L. Fischer, 1980, Gel Filtration Chromatography, Elsevier, Amsterdam
35. J. Grun et al, 1992, *BioPharm*, **5**(9) 22
36. Dove, G., G. Mitra, G. Roldan, M. Shearer, and M.-S. Cho, 1990, ACS Symposium Series #427, Chap. 14, American Chemical Society, Washington D.C.
37. G. Sofer and L.-E. Nystrom, 1991, Process Chromatography, A guide to Validation, Academic Press, San Diego
38. M. Weary and F. Pearson, 1988, *BioPharm*, **1**(4) 22
39. S.-C.Yan et al, 1984, *Anal. Biochem.*, **138** 137
40. J. Brewer and L. Soderberg, 1977, in Chromatography of Synthetic and Biological Polymers, (R. Rpton, ed.), Vol. 1, p.285, Ellis Horwood, Chichester
41. B.-L. Johansson and C. Ellström, 1985, *J. Chromatogr.*, **330** 360
42. J. Pedvelskis, 1993, personal communication
43. W. Kopaciewicz and F. Regnier, 1982, *Anal. Biochem.*, **126** 8
44. H. Engelhardt and D. Mathes, 1981, *Chromatographia*, **14** 325
45. D. Schmidt et al, 1980, *Anal. Chem.*, **52** 177
46. H. Engelhardt and M. Hearn, 1981, *J. Liquid. Chromatogr.*, **4** 1261
47. M. Hearn et al, 1980, *J. Liquid. Chromatogr.*, **3** 1549
48. E. Pfankoch et al, 1980, *J. Chromatogr. Sci.*, **18** 430
49. R. Frigon et al, 1983, *Anal. Chem.*, **55** 1349
50. M. Belew et al,1978, *J. Chromatogr.*, **147** 205
51. S. Aoyagi et al, 1982, *J. Chromatogr.*, **253** 133
52. T. Hashimoto et al, 1978, *J. Chromatogr.*, **160** 301
53. D. Eaker and J. Porath, 1967, *Sep. Sci.*, **2** 507
54. H. Barth, 1980. *J. Chromatogr. Sci.*, **18** 409
55. M. Janado et al, 1976, *J. Biochem.*, **79** 513

56. M. Janado et al, 1976, *J. Biochem.*, **80** 69
57. L. Politi et al, 1983, *J. Chromatogr.*, **267** 403
58. A. Galkin et al, 1984, *Anal. Biochem.*, **142** 252
59. T. Fujita et al, 1980, *J. Biochem.*, **87** 89
60. J.-C. Janson and J.-A. Jönsson, 1989, in Protein Purifiction; Principles, High resolution Methods, and Applications, (J.-C. Janson and L. Rydén, eds.), p.35, VCH, Cambridge
61. B. Gelotte, 1960, *J. Chromatogr.*, **3** 330
62. TosoHaas Technical Report #10
63. C. Middaugh and G. Litman, 1977, *J. Biol. Chem.*, **252** 8002
64. C. Tanford, 1968, *Adv. Protein Chem.*, **23** 121
65. W. Jencks, 1969, Catalysis in Chemistry and Enzymology, p.323, McGraw-Hill, New York
66. S. Timasheff and G. Fasman, 1969, Structure and Stability of Biological Macromolecules, Marcel Dekker, New York
67. P. Clezardin et al, 1986, *J. Chromatogr.*, **354** 425
68. P. Clezardin et al, 1986, *J. Chromatogr.*, **358** 209
69. J. Porath and N. Ui, 1964, *Biochim. Biophys. Acta*, **90** 324
70. U.-B. Hansson and E. Nilsson, 1973, *J. Immunol. Met.*, **2** 221
71. E. Kabat and M. Mayer, Experimental Immunochemistry, 2nd Ed., p. 264, Charles C. Thomas Publisher, Springfield
72. J. Porath and B. Olin, 1983, *Biochemistry*, **22** 1621
73. Y. Kato et al, 1981, *J. Chromatogr.*, **205** 185
74. J.-C. Janson and P. Hedman, 1982, in Advances in Biochemical Engineering (A. Fietcher, ed.) Vol. 6, p.43, Springer-Verlag, New York
75. P. Bristow and J. Knox, 1977, *Chromatographia*, **10** 279
76. L. Hagel, 1985, *J. Chromatogr.*, **324** 422
77. T. Andersson and L. Hagel, 1984, *Anal. Biochem.*, **141** 461
78. C. Middaugh et al, 1978, *Clin. Lab. Immunol.*, **1** 141
79. C. Middaugh et al, 1980, *J. Biochem.*, **255** 6532
80. C. Middaugh et al, 1978, *Proc. Nat. Acad. Sci. USA*, **75** 3440
81. Y. Kato et al, 1981, *J. Chromatogr.*, **206** 135
82. Y. Kato et al, 1981, *J. Chromatogr.*, **208** 71
83. —, 1991, Gel Filtration, Principles and Methods, 5th edition, Pharmacia, Uppsala
84. M. Carlsson et al, 1985, *J. Immunol. Met.*, **79** 89
85. B. Malm, 1987, *J. Immunol. Met.*, **104** 103
86. G. Perry et al, 1984, *Prep. Biochem.*,**14**(5) 431
87. S. Burchiel et al, 1984, *J. Immunol. Met.*, **69** 33
88. F.-M. Chen et al, 1988, *J. Chromatogr.*, **444** 153
89. A. Wichman and H. Borg, 1977, *Biochim. Biophys. Acta*, **490** 363
90. F. Franek, 1986, *Met. Enzymol.*, **121** 631

Chapter 4

Ion Exchange Chromatography

"Don't fight forces, use them!"
—R. Buckminster Fuller

Ion exchange chromatography (IEC) was first successfully applied to fractionation of plasma proteins in the mid–1950s.[1-3] By the 1970s it was firmly established as a standard method for purification of polyclonal antibodies.[4-10] This provided a foundation for its application to monoclonals and it evolved rapidly to become the dominant method in the field.[11] Advances in support media, including both chromatography and filtration formats, continue to expand its capabilities and flexibility.

Mechanism

Charged residues on protein surfaces include the side groups of amino acids, the α-amino and α-carboxyl termini of the polypeptide chains, and sialic acid residues on glycoproteins. These residues are amphoteric, making the sign and net charge on proteins a function of pH. (Table 4.1).[12-15] The pH at which a protein's positive charge balances its negative charge is its isoelectric point (pI). At pH values above their pIs, proteins become progressively electronegative. Below their pIs, they become progressively electropositive.

The traditional ion exchange model suggests that pI is the sole determinant of a protein's retention behavior.[16,17] A protein should bind to an anion exchanger at pH values above its pI and to a cation exchanger below its pI. Retention should diminish rapidly to zero on both as pH approaches the pI. Anion exchange selectivity should precisely mirror selectivity on cation exchange.

Table 4.1. pKs for charged amino acids. The altered ranges in proteins reflect the influence of local surface characteristics. Proximity of unlike charges elevates the pKs of positively charged residues and lowers the pKs of negative groups. Proximity of like charges or hydrophobic regions have the opposite effect. The protein-value for arginine is predicted. All others were measured experimentally.[12-15]

Residue	pK in amino acid	pK in protein
α-amino	8.8–10.8	6.8–7.9
arginyl	12.5	≥12
histidyl	6.0	6.4–7.4
lysyl	10.8	5.9–10.4
α-carboxyl	1.8–2.6	3.5–4.3
aspartyl	3.9	4.0–7.3
glutamyl	4.3	4.0–7.3

Rigorous studies show that less than 25% of proteins accurately fulfill the model's predictions.[16-23] Some bind a full pH unit below their pI on anion exchangers and a full unit above on cation exchangers. This indicates that positive and negative sites dominate different portions of the molecule. Others fail to bind a full unit above their pI on anion exchangers and/or a full unit below on cation exchangers. This indicates that negative and positive charges are mixed in the binding regions, partially neutralizing their respective interactions with the exchanger. Isoproteins with different primary sequences but identical pIs exhibit different retention properties, while among antibodies, differentially charged glycosylation isoforms of a single monoclonal usually elute together in a single peak.

These departures don't indicate that the model is fundamentally wrong, merely that it's too general. The most important qualification is the distinction between the net charge of the protein as a whole and the charge characteristics of its actual binding site.[19] The binding site may include one or more charged residues, even multiple areas of a protein, but usually not all of its charged residues. An important corollary is that the composition of the binding site at one pH may be very different from its composition at another (Figure 4.1). For example, a histidyl-rich site may dominate binding on a cation exchanger at pH 5.5. Retention of the same protein at pH 7.5 may rely on a lesser number of untitrated residues on the opposite side of the pro-

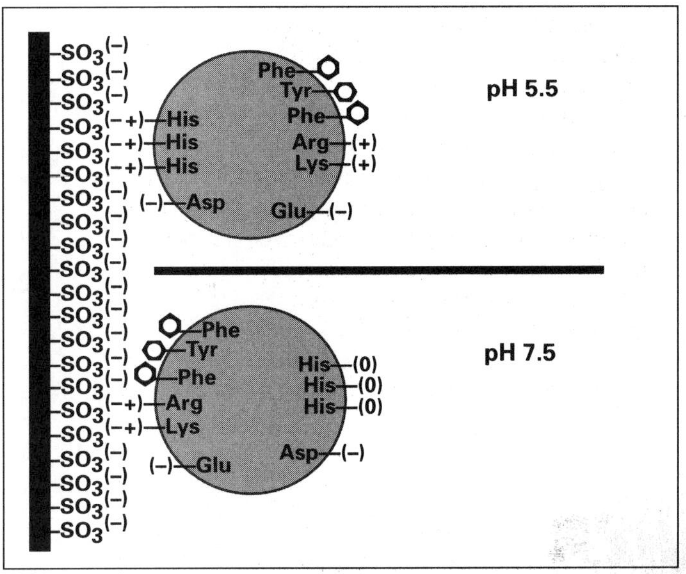

Figure 4.1. Preferential orientation on a cation exchanger as a function of pH. HIS: histidine. ASP: aspartic acid. PHE: phenylalanine. TYR: tyrosine. ARG: arginine. LYS: lysine. GLU: glutamic acid.

tein—residues that didn't even contribute to retention at pH 5.5. By extension, the respective binding sites of a given protein are likely to be different for anion and cation exchangers.

These qualifications have important ramifications for selectivity. The contaminant spectrum with which an antibody is associated on a cation exchanger is almost certain to be different from the spectrum on anion exchange. Both are worth exploring in their own rights. Even on the same exchanger, altering pH may significantly change the spectrum of contaminants coeluting with the product. This isn't to suggest that multistep purification processes will routinely benefit from having 2 steps on the same exchanger. It does mean that it's worth exploring a range of pH values to determine which provides the best complement to other candidate methods in the process.

Bound proteins can be eluted by altering pH, by addition of competing ions, or combinations of the 2. Each approach gives different selectivity. Achieving elution solely by altering pH at low ionic strength is the basis of a technique called chromatofocusing.[17,24-28] Proteins elute near their pIs. This turns out to be more of a liability than an asset with mono-

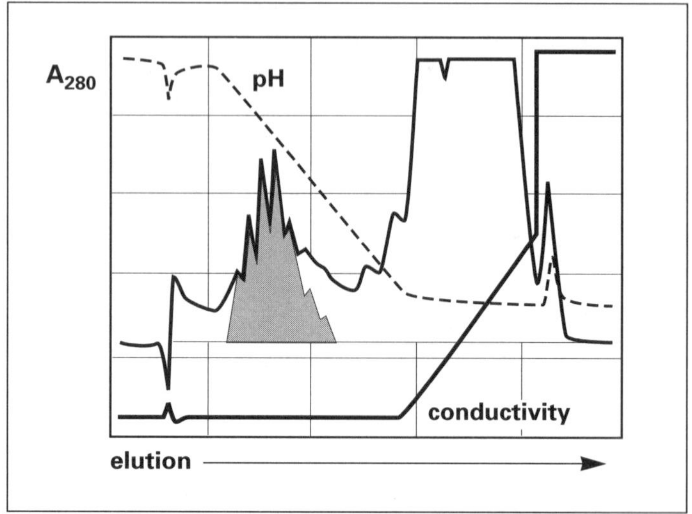

Figure 4.2. Analytical chromatofocusing profile of a mouse IgG_3 monoclonal in ascites. pH range 9–6. The column was eluted with a sodium chloride gradient following the pH elution. The monoclonal profile was determined with a separate injection of protein A-purified antibody.

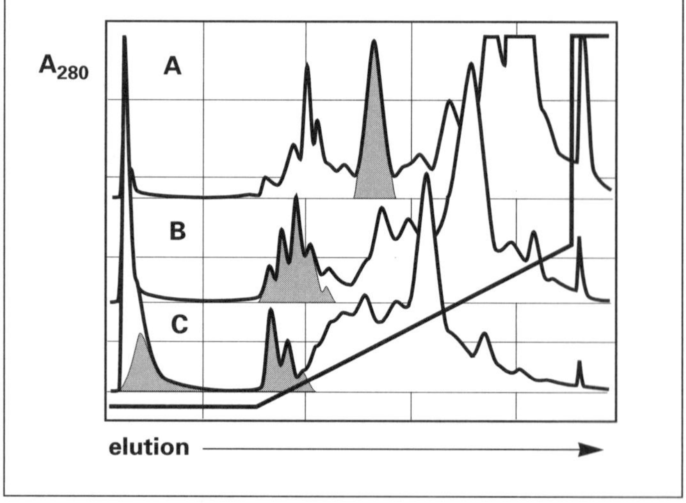

Figure 4.3. Salt-mediated glycoform fractionation as a function of pH. Profile A was run at pH 8.5, B at pH 7.0, and C at pH 5.5. The pI of this mouse IgG_1 was 6.4–6.7. Monoclonal profiles were determined by injecting protein A-purified antibody in a separate series of runs.

clonals because post-translational glycosylation isoforms elute in multiple peaks (Figure 4.2).[29,30] Glycoform fractionation is also observed with salt elution if the pH is near an antibody's pI (Figure 4.3). Monoclonals otherwise tend to elute in single peaks.

Besides controlling selectivity with gradient slope, it can be altered by employing different eluting salts. Ions of higher valencies exert more effect on retention than monovalent ions, but even ions of the same valency exert different effects.[16-20,31-35] Their relative abilities in this regard correlate with their respective rank-

Table 4.2. The Hofmeister series of lyotropic and chaotropic ions. The directions of the arrows indicate increasing effectivity of eluting ions.

Anions →								
SCN^-	ClO_4^-	NO_3^-	Br^-	Cl^-	COO^-	SO_4^{2-}	PO_4^{3-}	
← Cations								
Ba^{2+}	Ca^{2+}	Mg^{2+}	Li^+	Cs^+	Na^+	K^+	Rb^+	NH_4^+

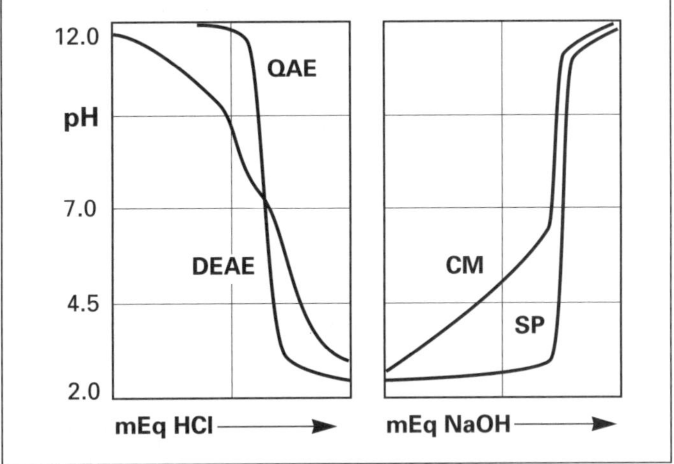

Figure 4.4. Titration curves for common ion exchange ligands. QAE indicates quarternary aminoethyl, DEAE indicates Diethylaminoethyl, CM indicates carboxymethy, and SP indicates sulfopropyl. The stability of the QAE and SP curves over a wide pH range demonstrates why they are called "strong" ion exchangers. This resistance to titration translates into level protein binding capacity over a wide pH range. It has no direct significance with respect to selectivity.

ings in the Hofmeister series (Table 4.2).[35,36] Allosteric interactions with specific ions may alter protein conformation sufficiently to redefine the composition of the binding site.[20,37-40]

On a practical level, it's not worthwhile to screen a wide range of eluting salts. Differences in displacing strengths only alter the virtual slope of an elution gradient, and allosteric effects are unpredictable.[17,40] Consequently, sodium chloride is used to the virtual exclusion of all other candidates.

Differences among ion exchangers also contribute to selectivity.[31] The titration characteristics of the ligand determine an exchanger's fundamental charge properties (Figure 4.4). Variations in the spacer arm contribute to the hydrophobicity of the support overall and are the primary determinant of ligand accessibility. This is why tentacle-type supports provide different selectivities than short-spacer supports—the ligand has 3-dimensional access to the proteins.[41] The support matrix itself exerts an influence by way of both its pore size distribution and surface hydrophobicity. All other features being equal, ion exchangers with stronger hydrophobicities bind a broader spectrum of proteins and bind them more strongly than less hydrophobic exchangers.

Attributes

IEC is applicable to all monoclonal antibodies, regardless of class, species of origin, or production method (Figure 4.5).[11,42-63] This is not to say that all antibodies are equally suited to preparative fractionation by both exchangers. Some IgGs are too basic to support high binding capacities on anion exchangers within reasonable pH limits (≤8.6). These particularly include IgG_3s. Acidic antibodies sometimes require excessively low pH (<4.5) for high capacity binding on cation exchangers.

purification performance

Ion exchangers provide 60–95% purity from raw production media. Some reports suggest that 1-step purifications are feasible, but this is neither representative nor realistic. These suggestions ignore the real-life requirements of commercial purification. High single-step purity requires narrow peak cutting and the reduced recovery is likely to be more costly than the addition of a supporting purification step. Such suggestions also fail to acknowledge that even the most robust methods are vulnerable to external process variations.

removal of nonspecific antibody

IEC supports 50–75% removal of host or media supplement-derived polyclonal antibody. Combined reduction by both exchangers often exceeds 90%.

DNA removal

IEC usually supports 3–5 logs removal of DNA. It binds strongly to anion exchangers by its negatively charged phosphate moieties.[42,61-65] The same groups cause it to be repelled and pass unretained through cation exchangers. An important exception to the DNA reduction capabilities of both methods occurs when DNA is complexed to the product. DNA is known to form stable charge complexes with proteins and has even been used as an "affinity" ligand for chromatographic purification of antibodies.[66-69] The antibodies most frequently and severely affected are IgMs and strongly basic IgGs. The low ionic strength

Figure 4.5. Paired anion and cation exchange elution characteristics of 17 monoclonal antibodies. Anion exchange results were obtained with 0.05M Tris, pH 8.6. Cation exchange results were obtained with 0.05M MES, pH 5.6.

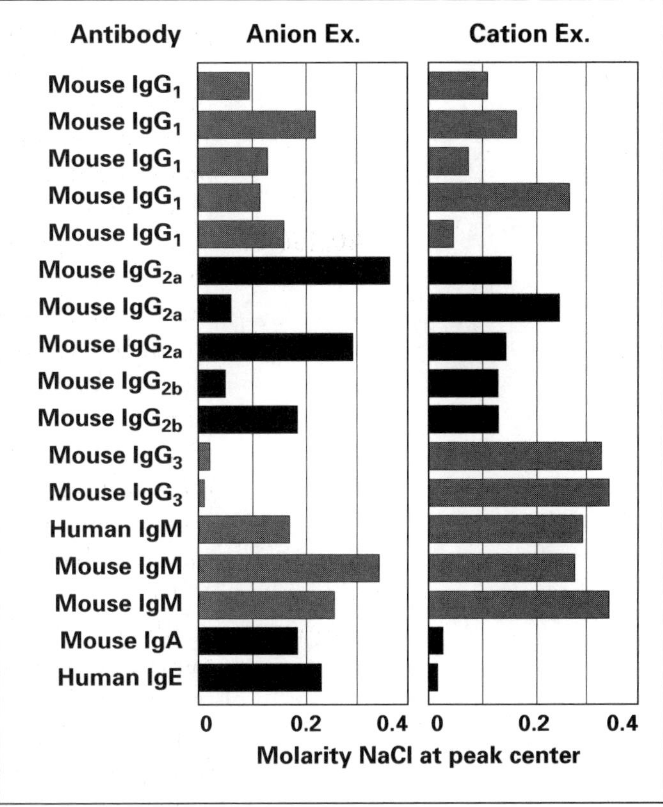

at which IEC is normally carried out favors such complexation. The product can transport high DNA levels all the way through such a process (Figure 4.6).[70]

An important side note: complexation also affects DNA assays. Dissociation and removal of interfering protein are necessary to obtain accurate measurements.[71,72]

virus removal

Anion exchange supports 3–6 logs clearance of viruses; cation exchange, somewhat less (Table 4.3).[43,44,61,62,64,65,73] Although the clearance capabilities of cation exchange are generally lower, its selectivity is complementary. Combining the 2 chemistries increases the net clearance factor, as does combining either or both of them with other methods. An important caution: design column sanitization procedures with sufficient viral-kill capability to ensure that viruses left on the column don't contaminate subsequent product lots.

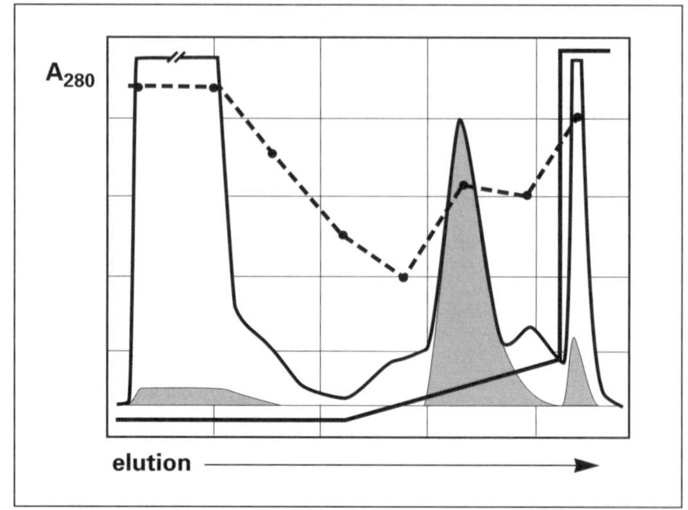

Figure 4.6. DNA transport and product loss due to ionic complexation with an IgM. This chromatogram illustrates a first-step cation exchange separation from DNA-contaminated cell culture supernatant. Antibody distribution is indicated by the shaded areas. DNA in pg/mL is indicated by the dashed line. From highest to lowest, the values range from ~10^{10} to ~10^{6}. See text for discussion.

endotoxin removal

Anion exchangers remove 2–5 logs of endotoxin.[42,61,62,64,65,74] At alkaline pH, phosphoryl and carboxyl moieties from the lipid A and core polysaccharide region bind strongly while the ethanolamine moieties are largely titrated and minimally influential.[75,76] Since most antibodies are weakly bound to anion exchangers, separation is typically good. Cation exchangers may eliminate up to 2 logs of endotoxin. At acidic pH, the influence of endotoxin negative charges is reduced by titration, but the ethanolamines exert their full positive charge. Since antibodies are among the stronger binding components on cation exchangers, they tend to coelute to some degree. Phenol red and other negatively charged media additives are largely removed by binding to anion exchangers, or by their repulsion from cation exchangers.[44,45,77-80]

mass recovery

Mass recovery for IgG monoclonals ranges from 80–95%, averaging about 90%. Lower estimates often result from failing to account for selective removal of nonspecific antibodies. IgM recoveries range from 60–90%, also averaging toward the higher side. Lower recoveries usually reflect process design rather than an inherent limitation of the technique. Inadequate binding conditions and excessively narrow pooling are both frequent causes. However, as with purification, DNA complexation is sometimes a problem.[70]

Table 4.3. Viral clearance by IEC. Best results from references cited in the text. All values given in $\log_{10}$. E-MULV: ecotropic murine leukemia virus. X-MULV: xenotropic murine leukemia virus. HBV: hepatitus B virus. HSV: Herpes simplex virus. Polio: polio virus.

Virus	Anion Ex.	Cation Ex.
E-MULV	5.7	2.8
X-MULV	2.0	2.8
HBV	5.0	—
HSV	5.1	3.4
Polio	1.6	—

losses due to DNA complextion

Product loss from DNA complexation in Figure 4.6 was attributed to either or both of 2 phenomena: DNA neutralization of positive charges on the protein, or DNA:monoclonal complexes large enough to be excluded from the pores of the media, but small enough to pass through the void volume. The loss of product in the 1.0M sodium chloride strip was attributed to complexes too large to penetrate the column being trapped at the inlet until they were dissociated by the elevated salt concentration. Product recovery from the main peak was less than 70%. When the DNA was dissociated and removed in a previous HIC step, cation exchange recovery increased to >90% with all the product eluting in the main peak.

Recovery of immunoreactivity per specific mass unit

IEC is regarded as a nondenaturing technique and activity recovery is usually quantitative. In fact, antibodies bound to ion exchangers appear to be more stable than they are under identical conditions in free solution. In one case, antibodies bound for several hours on a cation exchanger at pH 3.0 showed no activity loss, even though denaturation under these conditions in free solution is typically severe.[60] This is not to imply that all antibodies will survive this treatment without deleterious effect. Operating pH is best maintained within the range of pH 4.5–8.5.

Apparent activity recoveries greater than 100% result from failing to adjust titer measurements for removal of polyclonal contaminants. Activity recovery is often reduced somewhat for IgMs. This is more often encountered with cation exchange than with anion exchange. It relates to poor conformational stability at low pH and conductivity, and tends to be reflected in poor mass recoveries as well. IgMs often bind very strongly to one or both exchangers (Figure 4.5). This translates into the ability to fractionate them under more moderate pH conditions, improving mass and activity recovery at the same time.

high dynamic capacity

Most IEC media have dynamic capacities of at least 15mg per mL for IgG antibodies, and many exhibit double that or more. As with all adsorption methods, dynamic capacity is strongly affected by a range of factors, particularly including composition of the mobile phase, eluent velocity, and competition with contaminants.[81,82] Dynamic capacity under actual application conditions must be determined experimentally on an individual basis.

rapid throughput

IEC methods can be conducted rapidly. Most polymeric media support eluent velocities of at least 200cm/hr, allowing analytical runs to be conducted in ~30 minutes and preparative runs in 2–4 hours. The recent generation of chromatography media allows analytical runs to be completed in <5 minutes and preparative runs in <30. This encourages thorough process characterization, with large dividends during scale-up and validation.

resistance to harsh sanitizing conditions

Most IEC media withstand exposure to strong acids, bases, chaotropes, detergents, and organic solvents. This provides the basis for a wide range of column regeneration and sanitization methods. Expect to obtain several hundred runs from prepacked HPLC columns. The life of large-scale columns used for industrial separations varies according to the composition of the samples applied to it, and to the methods and frequency of cleaning and sanitization. With appropriate sample preparation and good column maintenance, preparative columns exhibit the same longevity as small HPLC columns.

Apply cleaning flow in the direction opposite that in which sample is applied. This prevents strong-binding contaminants from becoming distributed throughout the bed. Badly fouled columns may be unrecoverable. In this case, discard and replace the top cm of media at the inlet end of the column.

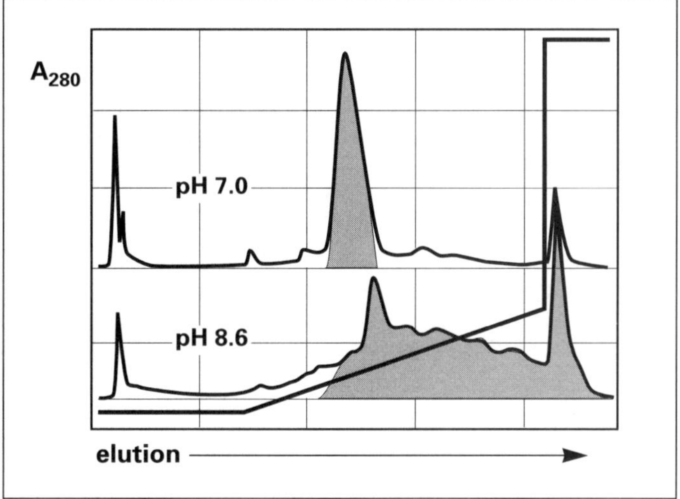

Figure 4.7. On-column denaturation of IgM as a function of pH. Sample: an unusually labile mouse IgM after preliminaty purification by size exclusion chromatography. The run at pH 7.0 was conducted in HEPES. The run at pH 8.6 was conducted in Tris.

Limitations

One of the most important and least recognized risks with IEC is the potential for inadvertent loss of glycosylation isoforms (Figure 4.3). Even when monoclonals elute in a single peak there is always a cline in glycoform distribution from the leading to the tailing side. Aggressive peak cutting deletes the boundary forms. This is a concern because glycosylation has been shown to be an important determinant of antibody titer, pharmacokinetics, and stability.[78-80,83-86] This does not mean that all changes in glycoform distribution have dire effects. It does mean that you should use IEF to evaluate pooling specifications, that you should avoid process specifications that alter glycoform distribution, and that you must validate any such alterations.

occasional product denaturation

A minority of monoclonal antibodies suffer denaturation during IEC. This is observed most often with IgMs. The antibody in Figure 4.7 eluted normally at pH 7.0, but at pH 8.6 the profile indicated that it had undergone massive conformational change. The eluted antibody was turbid upon collection and precipitated entirely within an hour. Antibody solubility and stability can be improved by the addition of glycine to IEC buffers.[87-89] Since it's zwitterionic, even saturated solutions don't interfere with binding.

antibody solubility limitations

Most antibodies are soluble at the high pH, low conductivity conditions required for loading anion

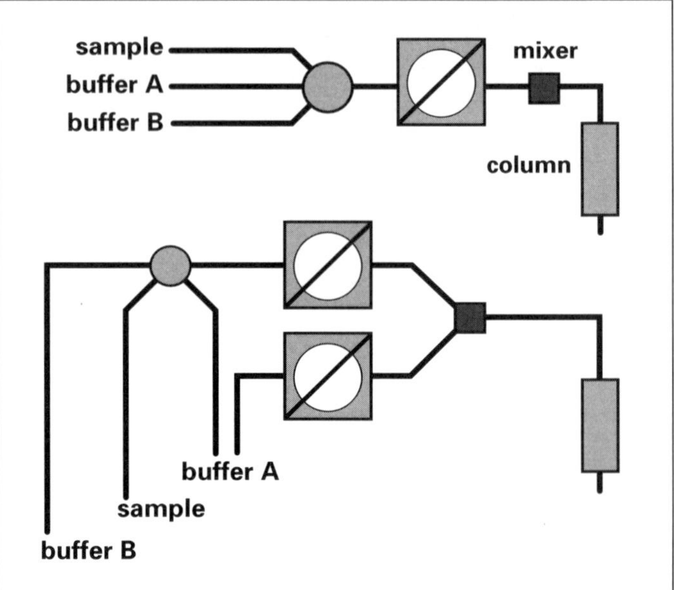

Figure 4.8. Sample application by on-line dilution. This technique can be practiced on any gradient chromatograph, but setup varies according to system configuration. Single-pump systems are the simplest. Dual-pump systems may require intermediate pump washes to protect the bound sample from premature contact with high-salt buffer. When doing on-line dilution for the first time, "walk through" the process thoroughly with a detailed plumbing diagram. Be especially watchful for dead-legs and test the setup with a small amount of sample before commiting large volumes.

exchangers. They can be batch equilibrated in advance and loaded directly onto the column. Very few are soluble under the low pH, low conductivity conditions required for cation exchange. This precludes advance equilibration since precipitates may cause clogging, not to mention reducing separation performance and recovery. Sample loading by on-line dilution provides a convenient method for managing such limitations.[90] Sample is diluted in the chromatograph by an equilibrating diluent buffer. Precolumn exposure of the product to precipitating conditions is too brief to permit problems to develop.

on-line dilution

Equilibration is best done in 2 steps. Just before application, titrate the sample to within ~0.5 pH units of the target value. The most consistent way of doing this is with a pretitrated 20x buffer concentrate that you add to the sample at 5% (v:v). Organic buffers are ideally suited because of their low conductivity. The next step is to identify the dilution factor. You will normally have defined the range of salt concentrations that support product retention during routine method development. Determine the proportion of diluent required to achieve target conductivity.

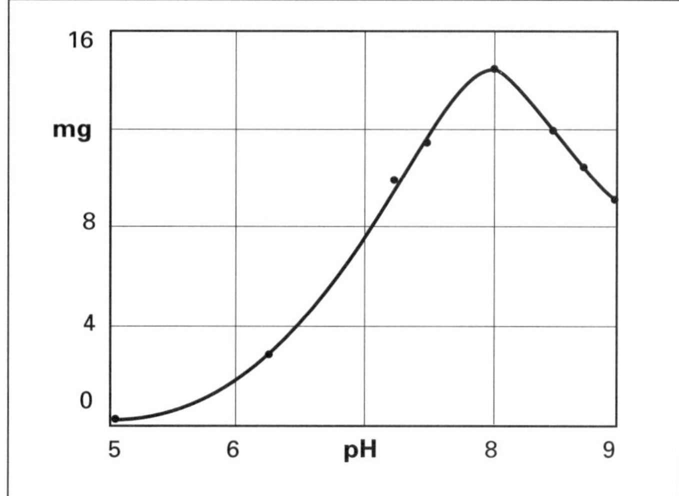

Figure 4.9. Proteolysis of IgG by copurified proteases as a function of pH. Purified antibody was dialysed at the indicated pH values and the buffer monitored for peptides. No proteases were added. Values shown are in mg peptide per g IgG. Data replotted from reference 91.

On-line dilution factors typically range from 3–5 but even factors up to 10 are preferable to abandoning a powerful fractionation technique. On-line dilution may support better process flow and economy than alternative methods, even when it's not strictly required to manage product solubility. Off-line equilibration steps require equipment, media, and documentation. Separate steps incur product losses. On-line dilution doesn't. Figure 4.8 illustrates on-line dilution configurations for 2 common gradient chromatograph formats.

risk of proteolysis

The potential for proteolysis in conjunction with IEC is important to be aware of. The high pH typically used for anion exchange is in the prime activity range for the trypsin proteases plasmin and kallikrein (Figure 4.9).[91] Two simple precautions are to limit the time from sample equilibration to column loading and to titrate samples to lower pH immediately after elution.

retention anomalies due to metal ion contamination

Antibodies bind metal ions avidly. This alters their net charge, reduces their pIs, and may significantly alter IEC retention characteristics.[92] Figure 4.10 illustrates a mouse IgG_1 that formed multiple charge variants by metal complexation.[93] The equilibrium was only partially responsive to EDTA. Even saturating levels were unable to drive the antibody exclusively to an apo-metallic form. Neither was exposure to excess metal ions able to drive it to a holo-form.

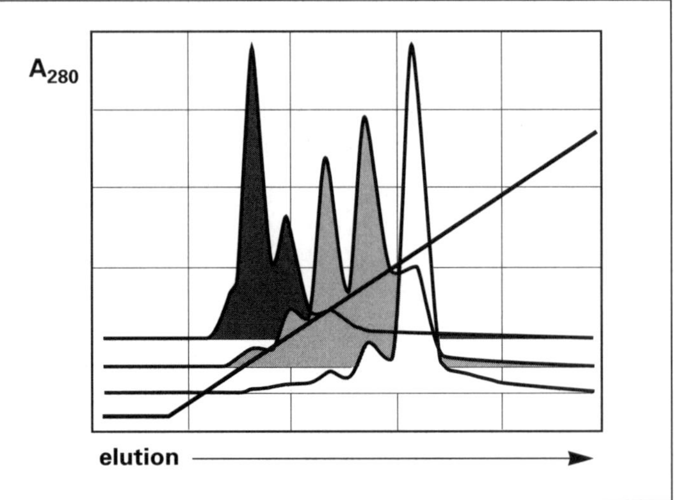

Figure 4.10. Dynamic heterogeneity of a mouse IgG_1 as a function of metal ion complexation. The dark gray profile illustrates elution of the purified antibody after treatment with 1mM zinc chloride. The light gray profile illustrates the zinc-treated antibody after buffer exchange into 1mM EDTA. The white profile illustrates the zinc-treated antibody after buffer exchange into 5mM EDTA. All separations were carried out with an anion exchanger at pH 8.6.

A mouse/human chimera exhibited only two forms. The metal-complexed form bound weakly to the anion exchanger while the apo-form coeluted with albumin. Both its unusually strong native retention and its metal-binding response were attributed to polycarboxy sites that formed coordination bonds with metal ions. This was discovered when the process was scaled up from a biocompatible development chromatograph to a stainless steel process system. Leached ions converted some of the product to the holo-form, which was lost during loading under conditions optimized for capture of the apo-form. Losses varied from 20–50%, depending on the condition of the chromatograph.

Besides the obvious consequences with separation behavior, bound metal ions destabilize proteins.[94,95] This highlights the importance of metal ion content as a process control variable. Use reagents that are controlled for metal contamination, keeping in mind that even part-per-million (ppm) concentrations can be significant. 1ppm nickel in a 1mg/mL solution of IgG equals a molar excess >50-fold. EDTA blocks formation of coordination complexes between metal ions and protein polycarboxy sites, but it can't prevent complexation with histidyl residues at alkaline pH.[96] This is a particular concern with IgG because of a highly conserved histidyl cluster at the juncture of the Cγ2 and Cγ3 domains.[97-99] Metal binding via this mechanism can be competitively blocked by inclusion of imidazole, histidine, or histamine in the process buffers.[96]

buffer capacity and pH control

Since pH is the variable most critical to selectivity, strong buffer capacity is essential to good process control. In this regard, it's important to keep in mind that stepped changes in the concentration of eluting ions directly affect local concentrations of titrating counterions. Stepping up the chloride concentration on anion exchangers displaces an equivalent concentration of ligand-associated hydroxide ions. Lacking adequate buffer capacity, this can elevate local pH as high as 12 and may denature your product.[100] Cation exchangers suffer acidification by hydronium ion displacement.

Evaluate buffering capacity by equilibrating your exchanger to start conditions, then continuously measuring effluent pH in conjunction with sodium chloride steps of 0.05*M*, 0.1*M*, 0.2*M*, and 0.4*M*. If this experiment reveals loss of pH control for the magnitude of gradient steps in your process, substitute a buffer with a pK closer to your operating pH. If the pK is already close, increase the buffer concentration. The alternative is to eliminate conductivity steps by eluting with a linear gradient.

high conductivity of inorganic buffers

A concern with buffers based on inorganic salts, whether anionic or cationic, is their large contribution to conductivity (Figure 4.11). In contrast, most organic buffers have very low conductivities. Conductivities are important because they affect binding capacity.

Figure 4.12 shows the relative effects of pH and conductivity on albumin capacity of an anion exchanger equilibrated with zwitterionic buffers.[101] Dynamic binding capacity was reduced >65% at the conductivity corresponding to 0.05*M* phosphate. This is an enormous sacrifice. The tendency is to compensate by reducing buffer concentration or using a more extreme pH. However, reducing buffer concentration reduces pH control and pH has limited ability to compensate. It would be necessary to alter pH by 1.5–2 units to regain the capacity sacrificed by using the inorganic buffer. Denaturation due to extreme pH becomes a concern.

Figure 4.11. Conductivities of 0.05M solutions of buffers commonly used for IEC. All values in Siemens/cm. A: Tris, pH 8.6. B: MES, pH 5.6 . C: HEPES, pH 7.0 . D: sodium acetate, pH 4.5 . E: sodium citrate, pH 4.5. F: sodium chloride, pH unadjusted. G: sodium phosphate, pH 8.0 . H: sodium phosphate, pH 7.0. Acetate and citrate buffers were made from free acid and titrated with sodium hydroxide to minimize conductivity.

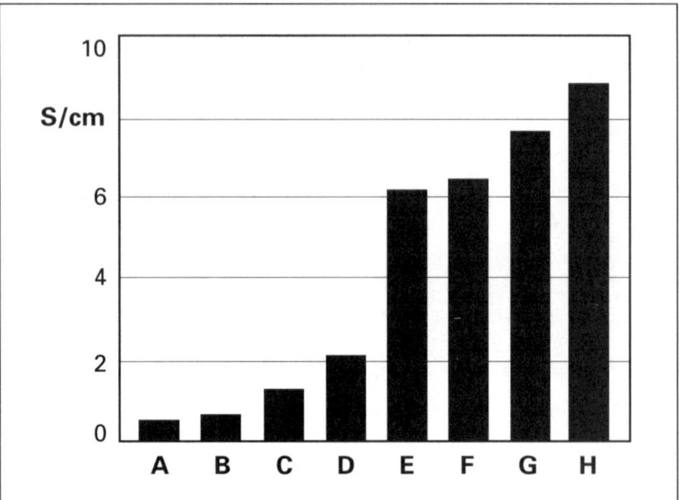

This raises the issues of material and validation costs. US Pharmacopeia (USP) buffers are often used because they are inexpexpensive and require minimal validation. However, besides mostly being anionic salts and providing very poor coverage of the range of pKs used in antibody purification, they are heavily contaminated with metals. The hidden process costs associated with these limitations can exceed the procurement and validation costs of their organic counterparts.

other buffer incompatibilities

Another concern with inorganic buffers is the ability of strongly basic antibodies to form stable crosslinks with polyvalent anionic buffers. These antibodies particularly include IgG_3s. In the presence of phosphate or citrate, at the low sodium chloride concentrations used for IEC, such antibody solutions become turbid. Filtering out the fine precipitate is not helpful because the equilibrium is re-established spontaneously upon removal. Precipitates can often be resolubilized by titrating them with sodium or potassium fluoride. The solution can then be buffer exchanged into a non-polyanionic buffer. Strong antibody binding to a cation exchanger is a warning sign.

column equilibration requirements

Whatever buffer you use, it is important to keep in mind that ion exchangers are themselves solid phase buffers. They require titration to equilibrate them to operating pH. Inadequate titration is a common cause

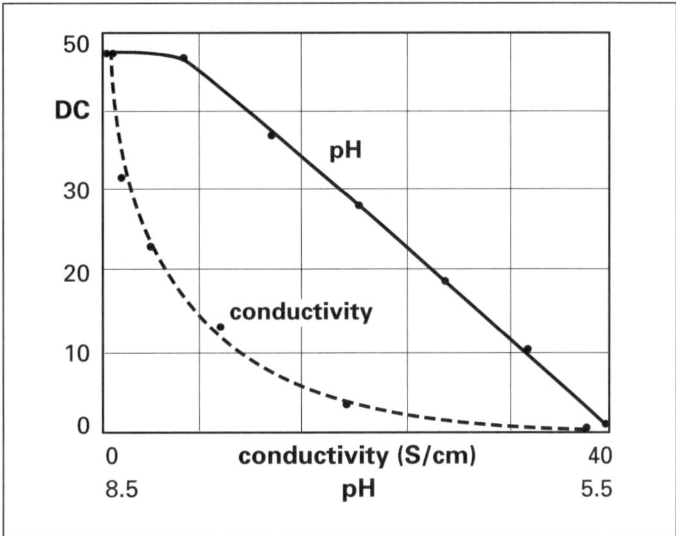

Figure 4.12. Dynamic capacity of bovine serum albumin on a strong anion exchanger as a function of conductivity and pH. DC indicates dynamic capacity in mg/mL of gel. Conductivity experiments were carried out at pH 8.5 with increments of sodium chloride. pH experiments were conducted with 0.05M Tris, HEPES, or MES in the absence of sodium chloride. Capacity plateaus at ~pH 8.0 because the albumin becomes fully titrated.

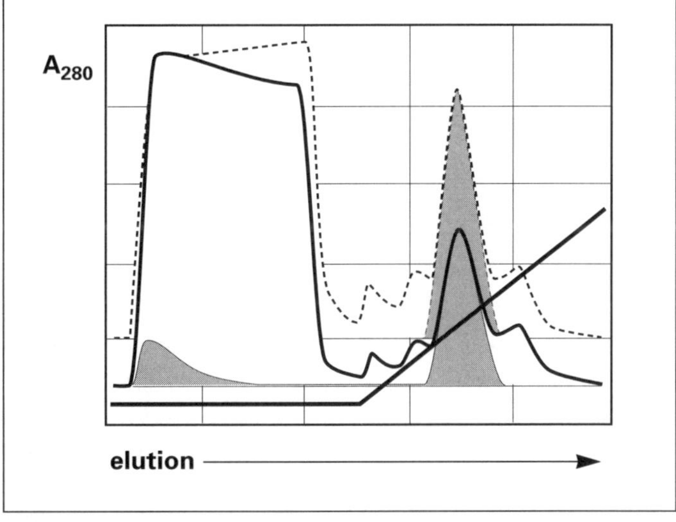

Figure 4.13. Scale-up failure due to inadequate pH titration of the column. Mouse IgG_1 ascites on a partially equilibrated cation exchanger. Note that the sample eventually completed column equilibration, after which the antibody bound normally. The dashed profile illustrates a reference chromatogram developed after complete equilibration. Contrast this failure mode with Figure 4.16.

of process failure. It's always desirable to minimize volumes of process solutions, but compromising process performance and reproducibility is an unjustifiable price for convenience (Figure 4.13). As a default, 10 column volumes (CV) of a 0.05M buffer within 0.5 pH units of its pK is usually adequate to achieve full equilibration. Evaluate lesser volumes thoroughly, especially if the buffer concentration is lower or the operating pH is farther from its pK.

vulnerability to fouling

Anion exchangers are notoriously vulnerable to fouling. Cell debris, DNA, endotoxins, lipoproteins, phospholipids, and phenol red are all negatively charged and bind strongly. This can be a serious problem if anion exchange is your first process step. However, it provides an opportunity to turn adversity to advantage. These foulants all bind under physiological conditions, while antibodies rarely do so. A broadly applicable foulant removal method exploiting this differential is described in appendix II.

temperature anomalies

IEC is little-affected by minor variations in temperature, but problems can be encountered when converting methods from room temperature to cold processing. pKs of charged protein residues vary with temperature just as they do for buffers. Conformational modifications may also occur that redefine composition of IEC binding sites. An occasional but severe example of the latter is cryoprecipitation: progressive antibody insolubility below 37°C. This is rare with IgGs but affects up to 20% of IgMs.[83-86]

Cryoprecipitation is driven mainly by polar interactions: both salt-bridging and hydrogen bonding. [102-104] It is relatively independent of pH from 5–10 and consequently affects both anion and cation exchangers.[105] Increased conductivity will suppress it in some cases, but generally at too high a salt concentration to support IEC binding. Nonionic hydrogen donor/acceptors such as 1.0*M* urea are the most effective blockers for use with IEC.[102] At this concentration there is no need for concern about denaturation.[106-108]

A final warning: cryoprecipitates are often transparent and nearly invisible, even on glass surfaces. This can lead to inadvertent product losses in a wide variety of circumstances. With antibodies that you identify to be cryoglobulins, make sure that pertinent SOPS clearly specify that samples must be raised to an appropriate temperature prior to transfer between vessels, and especially before conducting any separation process.

Method development

Selectivity screening can be accomplished quickly and effectively on small columns with micrograms of

Table 4.4. Buffers and screening conditions. BICINE, MES, and HEPES are all zwitterionic and can be used interchangeably with both exchangers.

Buffers

pH 5.5
A: 0.50M MES
B: A + 1.0M sodium chloride

pH 7.0
A: 0.05M HEPES
B: A + 1.0M sodium chloride

pH 8.5
A: 0.05M Tris or BICINE
B: A + 1.0M sodium chloride

Conditions

Equilibrate column: with 10CV buffer A
Inject: 2–5%CV unequilibrated sample
Wash: 2CV buffer A
Elute: in a 15CV linear gradient to 30% buffer B.
Strip: with a 5CV 100% buffer B.

sample. Samples in physiological environments can be applied to IEC columns without prior equilibration, so long as injection volume is maintained below 5% of the column volume (CV). A small amount of purified reference antibody is useful for identifying your monoclonal from complex elution profiles. This reduces reliance on laborious secondary analytical methods like electrophoresis or immunoassay.

screening conditions

Suggested screening buffers and conditions are listed in Table 4.4. Figure 4.14 illustrates the benefits of multiple pH screening. This monoclonal bound well to a cation exchanger at pH 5.6 and gave ~75% purity. Purity was closer to 90% at pH 7.0 and 95% at pH 8.5. It's uncommon for monoclonals to exhibit retention this strong, but the example is typical of the improved purification you can obtain when they do.

For pH values where the antibody binds, you will need to fractionate enough raw production media to obtain samples for further evaluation. Conduct experiments at qualified pH values under conditions identical to those in Table 4.5 except for sample loading. Load 2–5mg of antibody per mL of gel by on-line dilution, first equilibrating the sample pH off-line, then

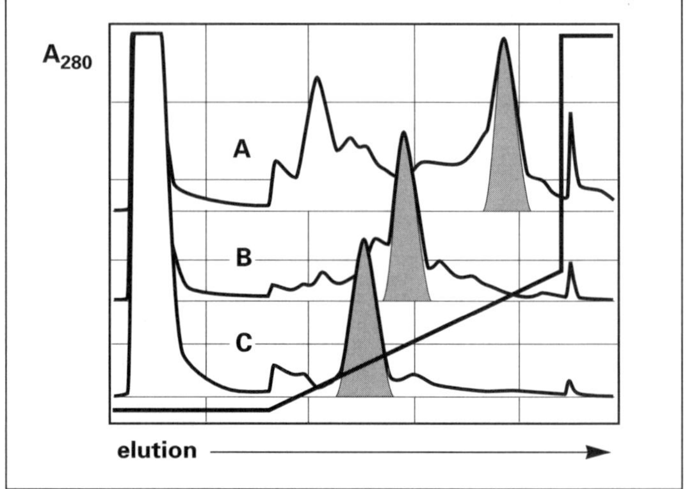

Figure 4.14. pH screening of a mouse IgG_1 by cation exchange. Sample: 20µL ascites on a 1mL high performance column.
A: pH 5.5
B: pH 7.0
C: pH 8.5

diluting on-line to reduce conductivity: 10% sample, 90% equilibration buffer.

performance evaluation

The context in which you evaluate the various run conditions is one of the most important aspects of process development. IEC is not a 1-step purification method. Consequently, the most important issue is how contaminants under a given set of screening conditions compare with the identity of contaminants in other candidate methods.[109] For example, if the antibody pool at pH 8.5 is 10% contaminated with a protein that also coelutes on HIC, but the pool at pH 7.0 is 30% contaminated with a protein that is unretained by HIC, then pH 7.0 is the better option. Information concerning percent purity is worthless except so far as the antibody in one method variant is accompanied by a lesser amount of the same contaminant(s) as another variant. Use percent purity to document the progress of a purification but ignore it as a design tool.

evaluating process combinations

With selectivity defined for IEC and other process candidates, you can identify the combinations that are likely to provide you with the best overall purification. Look for the smallest subset of methods in which no contaminants are shared.[109] You can also begin to give some thought to method sequencing. Anion exchange is a poor candidate for the initial process step because IgGs are among the weakest binding proteins.

Capacity is therefore consumed by the contaminants. The same contaminants are mostly unretained by cation exchangers, leaving maximum binding capacity for the product.

elution conditions and format

Once the method sequence is established, it defines the buffer and contaminant composition of the sample throughout the process. At this point you can optimize elution conditions. Elution can be conducted in either step or linear gradient formats. If you are purifying a product for in vitro application and your aim is to have no more than a 2-method process, you may need to use linear gradient elution.

Linear gradients also support better reproducibility. Most process variations have the effect of either weakening or strengthening retention. Such variations may alter virtual gradient slope or produce a modest frameshift, but so long as the product elutes near the middle and gradient amplitude exceeds the range of external variation, the relationships among eluting proteins are relatively unaffected.[110] This can be a valuable asset in the often less-controlled circumstances of diagnostic manufacturing. The only real disadvantage is that linear gradients dilute the product.

Process variables are typically better characterized and controlled in manufacture of injectables, and processes usually comprise 3–4 purification methods. This makes step gradients a lesser sacrifice in terms of purification performance and reproducibility. The logistics of dealing with large process volumes further favor the sample-concentrating ability of step gradients. This is not to suggest that antibodies for in vitro applications should categorically be purified with linear gradients, nor injectables exclusively with steps. You need to evaluate both and pick the format that best serves your needs.

column format

Choice of gradient format affects choice of column format and dimensions. Radial flow columns and membrane cartridges are good for step gradient applications, but their resolving capability with linear gradients compares poorly with axial flow columns. This is mainly due to inferior flow distribution. For axial flow

Table 4.5. Determination of dynamic capacity. This protocol is designed for experiments with pure proteins. However, other than the need for secondary product detection, the mechanics are the same for all samples. The correction factor described in steps 5 and 6 is important for studying phenomena that may have subtle influences on capacity but it can often be ignored in preparative applications. Measure it once and decide for yourself whether its worth the extra effort.

1. Put the column off-line
2. Flush the system with protein feed until the monitor signal plateaus
3. Make a chart mark coincident with putting the column in-line
4. Make a chart mark where the absorbance reaches 5% of the plateau value
5. Determine the correction factor for protein still in the system
   a. equilibrate the column to nonretaining conditions (ex., high salt)
   b. repeat steps 1–3, still under nonretaining conditions
   c. measure the volume of eluent from the chart mark to the point where protein is first detected at the monitor
6. Subtract the correction factor from the volume of eluent between the marks for steps 3 and 4.
7. Multiply the remaining volume times the protein concentration in the feedstream.

columns, a 5cm bed height provides 80% of the resolution of a 30 cm column.[111] Column beds as shallow as 2 cm have been used successfully in some preparative applications, but their performance relies heavily on the quality of flow distribution at the inlet. The risk of channeling is also very high.[112] Bed heights of 5–15cm are a good compromise.

Capacity

The last major process variable is dynamic binding capacity. A basic method and reference profile are provided in Table 4.5 and Figure 4.15. By the time you get to the last step of a process you may be able to monitor breakthrough from the UV absorbance profile. With the initial process step you'll need a secondary analytical method able to discriminate low breakthrough concentrations of product from overwhelming contaminant loads. Protein A and protein G have been employed effectively for this purpose.[113-116] Anti-light chain ligands can be used for non-IgGs.[117,118] All of these ligands capture nonspecific antibody, but so long as its proportion to

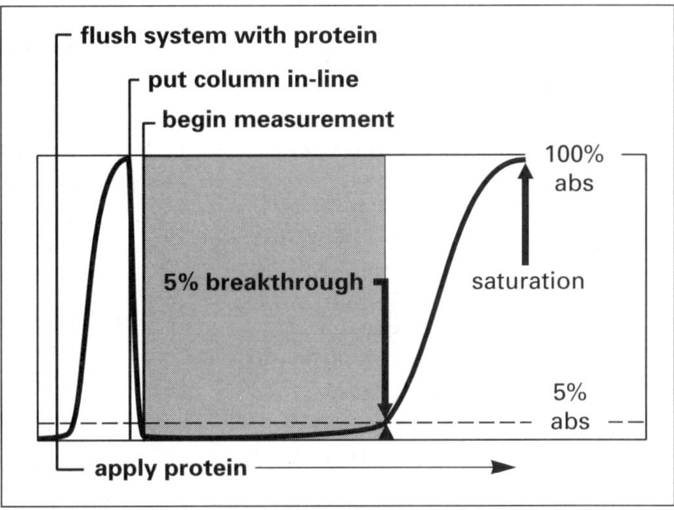

Figure 4.15. Determination of dynamic capacity. Refer to Table 4.5 for protocol. "abs" refers to spectrophotometric absorbance. The shaded area represents the amount of protein bound up to the point of 5% breakthrough.

the product is modest, it isn't a problem. It causes underestimation of capacity, which results in more conservative process specifications.

building in safety margins

Once you know dynamic capacity then you need to run your elution at that load and determine if separation performance is adequate. Reduce the load or fine-tune conditions as necessary to achieve the separation performance you require. How you set your final loading specifications depends on the level of variation you expect the scaled-up process to experience. If sample composition and process variables are rigidly controlled, you may be able to achieve good reproducibility with the specifications set at 95% of the capacity limit. If sample composition is variable and other parameters are poorly controlled, then reproducibility will be improved by setting the specification at a lower proportion: 90%, or even 80%. If you lack a good estimate of process variability, then set specifications as if it's going to be high. The expense of underloading a column is trivial in comparison to process failure.

scale-up issues

Most scale-up problems result from inadequate conditions for binding, either due to inadequate column equilibration, inadequate sample equilibration, or inappropriate loading conditions (Figures 4.13 and 4.16). Buffers are a surprisingly frequent source of scale-up problems, as a result of being prepared differently by

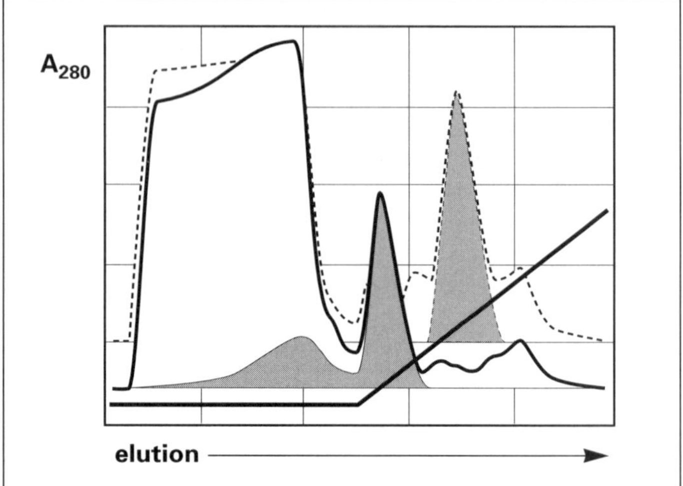

Figure 4.16. Scale-up failure from inadequate binding conditions or inadequate sample equilibration. Results like these usually result from a minor excess of salt in the binding buffer, in the sample, or in both. The dashed profile illustrates a reference chromatogram obtained under appropriate conditions. Note the difference in the elution position of the antibody compared to the failure mode illustrated in Figure 4.13.

different groups. Differences in raw materials are one cause. Development may use Tris base and titrate their buffers with hydrochloric acid. Manufacturing may use Tris-chloride and titrate with sodium hydroxide. The conductivity of the latter is significantly higher.

Conductivity discrepancies also arise from overshooting pH during formulation and correcting by back-titration. Backtitration elevates conductivity and should be strictly forbidden in IEC buffer SOPs. If you overshoot, discard the buffer and start over. The cost of materials is negligible compared to the potential impact on process performance. It's also important to recognize the effects of conductivity on pH. Some users make up twice the volume of starting buffer they need, adjust the pH, and add salt to half of it. The elevated conductivity drives down pH. Buffers must be titrated individually to ensure proper pH control.

Preparing large volumes of buffers in Manufacturing involves constraints that may be unapparent at bench scale. Process developers need to work with Manufacturing to accommodate these constraints. Development buffers should be prepared according to Manufacturing conventions, with Manufacturing approved materials, and with pH and conductivity meters on the Manufacturing calibration and maintenance program. Detailed SOPs are essential and periodic training should be conducted to ensure that all persons are specifically and fully informed as to how seemingly minor transgressions can affect downstream processing.

In perspective

The features that made IEC the dominant technique for antibody purification seem likely to maintain its ranking. It is broadly applicable, flexible, lends itself to systematic method development, and imposes few serious compromises. The fact that it can be exploited in 2 charge modes doubles its utility.

Reflecting the faith of suppliers, it has been the preferred platform for several processing innovations over the last few years. One of the most important has been the emergence of media able to support very high flow rates.[119-121] These mainly include chromatography media, but advances in filtration formats show increasing promise as well. High speed processing encourages more thorough process development and translates into better processes. Speed also translates into more efficient utilization of expensive equipment, manufacturing space, and labor.

Another important area of progress has been development of media for direct capture of product from crude feedstreams. Fluidized bed technologies allow debris and contaminants to pass between suspended beads without affecting product adsorption.[122] CellThru BigBead technologies take a different approach but achieve the same effect by allowing particulates to pass through the interstices between the oversized beads.[123] Neither of these technologies offer the same level of separation performance as conventional media but that's not their goal. By avoiding the need for preliminary clarification and concentration of cell culture supernatants they avoid significant sources of product loss, reduce overall costs by process simplification, and still provide a significant level of purification.

purification of in vitro products

Two-step combinations of IEC with HIC provide outstanding purification of monoclonals for in vitro applications. Cation exchange followed by HIC is especially effective, not only in terms of purification performance, but also because it doesn't require an intermediate buffer exchange step. High-flow media allow complete purifications to be conducted within 1–2 hours. Two-step IMAC/cation exchange procedures provide similar purification performance and process economy.

Anion/cation exchange combinations have been conducted successfully, but provide poorer performance on average than IEC combined with non-charge-based methods. Combinations of IEC with SEC work reasonably well for IgMs but less so for IgGs. Two-step purifications of salt or polyethylene glycol precipitation with IEC are still widely used but inferior to exclusively chromatographic combinations.

purification of in vitro products

IEC also provides a solid foundation for purification of in vivo products. Combinations of both exchangers with HIC, or either one with IMAC and HIC, support remarkable purification performance and process economy. In cases where a fourth step is required, SEC can provide a valuable polishing step, and concurrently streamline process flow.

Recommended reading

For a wide ranging general review of IEC consult the article by Karlsson.[112] Detailed discussion of protein charge distribution effects is found in references 19-21. References 11 and 42–63 describe ion exchange fractionation of monoclonal antibodies. Reference 124 describes IEC retention behavior of polyclonal antibodies from several mammalian species.

References

1. H. Sober and E. Peterson, 1954, *J. Am. Chem. Soc.*, **76** 1711
2. E. Peterson and H. Sober, 1956, *J. Am. Chem. Soc.*, **78** 751
3. H. Sober et al, 1956, *J. Am. Chem. Soc.*, **78** 756
4. H. Levy and H. Sober, 1960, *Proc. Soc. Exp. Biol. Med.*, **103** 250
5. B. Gelotte et al, 1962, *Arch. Biochem. Biophys. Suppl.*, **1** 319
6. J. Baumstark et al, 1964, *Arch. Biochem. Biophys.*, **108** 514
7. M. Joustra and H. Lundgren, 1969, in Protides of the Biological Fluids, (H. Peeters, ed.) Vol. 17, p. 511, Pergammon Press, Oxford.
8. H. Bjorling, 1972, *Vox Sang.*, **23** 18
9. A. Web, 1972, *Vox Sang.*, **23** 729
10. J. Fahey and E. Terry, 1978, in Handbook of Experimental Immunology, (D. Weir, ed.) Vol. 1, p. 8.1, Blackwell, London

11. J. Svasti and C. Milstein, 1972, *Biochem. J.*, **126** 837
12. C. Tanford, 1968, *Adv. Protein Chem.*, **23** 1
13. C. Cantor and P. Schimmel, 1980, Biophysical Chemistry, Vol. 1, Freeman, New York
14. R. Creighton, 1983, Proteins; Structure and Molecular Principles, Freeman, New York
15. D. Schmidt and F. Westheimer, 1971, *Biochemistry*, **10** 1249
16. —, 1980, Ion Exchange Chromatography: Principles and Methods, Pharmacia, Uppsala
17. —, 1991, FPLC Ion Exchange and Chromatofocusing, Principles and Methods, Pharmacia, Uppsala
18. L. Fägerstam et al, 1982, Protides of the Biological Fluids, Vol. 30, Pergamon Press, Oxford.
19. F. Regnier, 1987, *Science*, **238** 319
20. W. Kopaciewicz et al, 1983, *J. Chromatogr.*, **266** 3
21. J. Fausnaugh-Pollitt et al, 1988, *J. Chromatogr.*, **443** 221
22. R. Scopes and E. Elgar, 1979, *FEBS Lett.*, **106** 239
23. D. Brautigan et al, 1978, *J. Biol. Chem.*, **253** 130
24. L. Sluyterman and J. Wijdenes, 1977, in Proceedings of the International Symposium on Electrofocusing and Isotachophoresis, (R. Radola and J. Grasslin, eds.), p. 463, Walter de Gruyter, Berlin
25. L. Sluyterman and O. Elgermsa, 1978, *J. Chromatogr.*, **150** 17
26. L. Sluyterman and J. Wijdenes, 1978, *J. Chromatogr.*, **150** 31
27. L. Sluyterman and J. Wijdenes, 1981, *J. Chromatogr.*, **206** 429
28. L. Sluyterman and J. Wijdenes, 1981, *J. Chromatogr.*, **206** 441
29. A. Jungbauer et al, 1990, *J. Chromatogr.*, **512** 157
30. O. Kaltenbrunner et al, 1993, *J. Chromatogr.*, **639** 41
31. W. Kopaciewicz and F. Regnier, 1983, *Anal. Biochem.*, **133** 251
32. R. Drager and F. Regnier, 1986, *J. Chromatogr.*, **359** 147
33. K. Gooding and M. Schmuck, 1983, *J. Chromatogr.*, **266** 633
34. K. Gooding and M. Schmuck, 1984, *J. Chromatogr.*, **296** 321
35. R. Chicz and F. Regnier, 1990, *Met. Enzymol.*, **182** 392
36. F. Hofmeister, 1888, *Arch. Exp. Pathol. Pharmakol.*, **24** 247
37. W. Melander et al, 1984, *J. Chromatogr.*, **317** 67
38. J. Rosengren et al, 1975, *Biochim. Biophys. Acta*, **412** 51
39. H.-L. Wu, et al, 1986, *J. Chromatogr.*, **371** 3
40. L. Söderberg, 1980, in Protides of the Biological Fluids, (H. Peeters, ed.) Vol. 30, Pergammon Press, Oxford
41. —1994, Tentacle Ion Exchange Chromatography Handbook, E.M. Separations Technology, Gibbstown, NJ USA

42. G. Dove et al, 1990, ACS Symposium Series No. 427, (M. Ladisch et al, eds.), p. 194, American Chemical Society, Washington DC
43. S. Johnson-Camp, 1995, Clearance of Murine Leukemia Virus from a Chimeric Monoclonal Antibody using Ion Exchange Chromatography, poster, Prep-Tech '95, E. Rutherford, NJ USA, reprint: Bio-Rad Laboratories bulletin #1985, Hercules, CA USA
44. M. Gemski et al, 1985, *BioTechniques*, **3** 378
45. D. Nau, 1988, in Characterization and Analysis of Antibodies and Antibody Preparations, (D. Nau, ed.), Marcel Dekker, New York
46. A. Jehanli and D. Hough, 1981, *J. Immunol. Met.*, **44** 199
47. S. Ikeyama et al, 1986, *Molec. Immunol.*, **23** 159
48. P. Gallo et al, 1987, *J. Chromatogr.*, **416** 53
49. S. Burchiel et al, 1984, *J. Immunol. Met.*, **69** 33
50. A. Ross et al, 1987, *J. Immunol. Met.*, **102** 227
51. F. Jungbauer et al, 1987, *J. Chromatogr.*, **397** 313
52. P. Gavit et al, 1992, *BioPharm*, **1** 28
53. B. Pavlu et al, 1986, *J. Chromatogr.*, **359** 449
54. P. Clezardin et al, 1986, *J. Chromatogr.*, **358** 209
55. P. Clezardin et al, 1986, *J. Chromatogr.*, **354** 425
56. P. Clezardin et al, 1985, *J. Chromatogr.*, **319** 67
57. B. Moellering and C. Prior, 1990, *BioPharm*, **1** 34
58. J. Deschamps et al, 1985, *Anal. Biochem.*, **147** 451
59. M. Fitchum et al, 1994, Purification of a Human Monoclonal Antibody from Bioreactor Supernatant Using a Combination of Cation and Anion Exchange Chromatography, poster, International Symposium on Preparative Chromatography, Washington D.C, reprint: Bio-Rad Laboratories, bulletin 1917 US/EG Rev. B, Hercules, CA USA
60. M. Carlsson et al, 1985, *J. Immunol. Met.*, **79** 89
61. V. Garg, 1987, Use of Preparative HPLC in Large-Scale Purification of Therapeutic Grade Proteins from Mammalian Cell Culture, poster, 7th International Symposium on HPLC of Proteins, Peptides and Polynucleotides, Washington, D.C.
62. D. Nau, 1988, in The Role of HPLC in Biotechnology, (W. Hancock, ed.), John Wiley and Sons, New York
63. H. Levine, 1996, The Use of Membrane Adsorbers for Purification of Monoclonal Antibodies, slide presentation, The Waterside Monoclonal Conference, Norfolk, VA
64. T. Shibitani, et al, 1983, *Thromb. Hemostasis.*, **49** 91
65. W. Dembinski et al, 1983, Interferon Science Memorandum, **Jan./Feb** 6
66. M. Abdullah et al, 1985, *J. Chromatogr.*, **347** 129
67. R. Steiner, 1953, *Arch. Biochem. Biophys.*, **47** 56
68. R. Steiner, 1953, *Arch. Biochem. Biophys.*, **46** 291
69. D. Nau, 1990, *BioChromatography*, **5** 62

70. P. Gagnon, 1996, Special Weapons and Tactics for Removal of Product-Bound DNA, slide presentation, BioEast '96, Washington D.C.
71. M. Ishizawa et al, 1991, *Nucl. Acid Res.*, **19** 5792
72. —1992, A Phenol-Free DNA Extraction Method, Threshold Application Note 5, Molecular Devices, Sunnyvale, CA USA
73. J. Grun et al, 1992, *BioPharm*, **5**(9) 22
74. G. Sofer and L.-E. Nystrom, 1991, Process Chromatography, A guide to Validation, Academic Press, San Diego
75. M. Weary and F. Pearson, 1988, *BioPharm*, **1**(4) 22
76. R. Grabner, 1975, Process for removing pyrogenic material from aqueous solutions, U.S. Patent 3,897,309
77. F.-M. Chen et al, 1988, *J. Chromatogr.*, **444** 153
78. R. Parehk, 1992, *Biotech, Lab.*, **11** 61
79. Y. Kagawa, 1988, *J. Biol. Chem.*, **263** 508
80. M. Malaise, 1987, *Clin. Immunol. Immunopathol.*, **45** 1
81. J.-C. Janson and P. Hedman, 1987, *Biotechnol. Progr.*, **3**(1) 9
82. P. Gagnon and E. Grund, 1996, *BioPharm*, **9**(3) 34
83. B. Maiorella et al, 1993, *Bio/Technology*, **11**(3) 387
84. T. Monica et al, 1993, *Bio/Technology*, **11**(4) 512
85. C. Goochee et al, 1993, *Bio/Technology*, **8**(5) 421
86. C. Goochee et al, 1992, in Frontiers in Bioprocessing II, (P. Todd et al, eds.) p.199, American Chemical Society, Washington D.C.
87. J. Porath and N. Ui, 1964, *Biochim. Biophys. Acta.*, **90** 324
88. U.-B. Hansson and E. Nilsson, 1973, *J. Immunol. Met.*, **2** 221
89. E. Kabat and M. Mayer, 1966, Experimental Immunochemistry, 2nd Ed., p. 264, Charles C. Thomas Publisher, Springield.
90. P. Gagnon et al, 1995, Developing Sample Application Conditions for Large Scale Ion Exchange, poster, IBC Conference on Purification of Monoclonal Antibodies, San Francisco , reprint <http://www.validated.com/library.html>
91. B. Robert and R. Bockman, 1967, *Biochem. J.*, **102** 554
92. F. Gurd and P. Wilcox, 1956, *Adv. Protein Chem.*, **11** 312
93. —Anti-human IgG_4, clone RJ3, CalTag Laboratories, S. San Francisco, CA USA
94. T. Arakawa and S. Timasheff, 1982, *Biochemistry*, **21** 6545
95. T. Arakawa and S. Timasheff, 1984, *Biochemistry*, **23** 5912
96. J. Porath and B. Olin, 1983, *Biochemistry*, **22** 1621
97. E. Kabat et al, 1987, Sequences of Proteins of Immunological Interest, US Dept. of Health and Human Services, Public Health Service, National Institute of Health
98. J. Deisenhofer et al, 1978, *Hoppe-Seylar's Z. Physiol. Chem.*, **359** 975
99. J. Hale and D. Beidler, 1994, *Anal. Biochem.*, **222** 29
100. —, 1984, A Practical Guide to Ion Exchange, LKB, Bromma

101. P. Gagnon, 1996, *Valid. Biosys.*, **1**(2) 1 <http://www.validated.com/library.html>
102. C. Middaugh and G. Litman, 1977, *J. Biol. Chem.*, **252** 8002
103. C. Middaugh et al, 1978, *Clin. Lab. Immunol.*, **1** 141
104. C. Middaugh et al, 1980, *J. Biol. Chem.*, **255** 6532
105. C. Middaugh et al, 1978, *Proc. Nat. Acad. Sci.*, **75** 3440
106. C. Tanford, 1968, *Adv. Protein Chem.*, **23** 121
107. W. Jencks, 1969, Catalysis in Chemistry and Enzymology, p. 323, McGraw-Hill, New York
108. S. Timaseff and G. Fasman, 1969, Structure and Stability of Biological Macromolecules, Marcel Dekker, New York
109. P. Gagnon et al, 1993, *LC-GC*, **11** 26
110. P. Gagnon et al, 1995, *BioPharm*, **8**(4) 36
111. G. Vanacek and F. Regnier, 1980., *Anal Biochem.*, **109** 345
112. E. Karlsson et al, 1989, in Protein Purification: Principles, High Resolution Methods, and Applications, (J.-C. Janson and L. Rydén, eds), p. 107, VCH, New York
113. B. Compton et al, 1989, *Anal. Chem.*, **61** 1314
114. G. Blank and D. Vetterlein, 1990, *Anal. Biochem.*, **190** 317
115. ProAna Mabs, Hyclone Labs, Logan, UT USA
116. S. Fulton et al, 1991, *Biotechniques*, **11**(2) 226
117. P. Grandics, 1994, *Am. Biotech. Lab.*, **Jun.** 12
118. QuickMabs, Sterogene Bioseparations, Carlsbad, CA
119. POROS, PerSeptive Biosystems, Cambridge, MA
120. SOURCE, Pharmacia Biotech, Uppsala
121. Hyper-D, BioSepra S.A, France
122. STREAMLINE, Pharmacia Biotech, Uppsala
123. CellThru BigBeads, Sterogene Bioseparations, Carlsbad, CA
124. Y.-B. Yang and K. Harrison, 1995, Influence of Column Types and Chromatographic Conditions in the IEX Chromatography of Antibodies, poster, 15th International Symposium on HPLC of Proteins, Peptides, and Polynucleotides, Boston, rerint: Vydac, Hesperia, CA

Chapter 5

Hydroxyapatite Chromatography

"Look beneath the surface; let not the several quality of a thing nor its worth escape thee."
—Marcus Aurelius

Hydroxyapatite chromatography (HAC) was introduced in 1956, but despite numerous publications describing its merits for antibody purification, it has failed to attain the popularity of ion exchange.[1] A number of reasons account for its neglect, most of which are artificial and the remainder obsolete. In practice, it offers a unique assemblage of process characteristics that merit serious evaluation.

Mechanism

Unlike adsorptive chemistries where a reactive ligand is affixed to a "neutral" matrix, hydroxyapatite is both the ligand and the matrix. Its formula is $Ca_{10}(PO_4)_6(OH)_2$.[2] The functional groups comprise positively charged pairs of crystal calcium ions (C-sites) and clusters of 6 negatively charged oxygen atoms associated with triplets of crystal phosphates (P-sites). C-sites, P-sites, and hydroxyls are distributed in a fixed pattern on the crystal surface.[2-4]

As indicated by the formula, the Ca:P ratio should be 1.67. Hydroxyapatite synthesized by the traditional Tiselius method or adjustments thereto exhibit ratios of 1.50–1.55, indicating excess phosphate in the structure.[1,5-8] This leads to formation of unstable rectangular plate-shaped crystals with poor flow, poor pressure resistance, and poor stability characteristics. Recently developed synthesis methods yield hexagonal-cross section columnar crystals with the ideal Ca:P ratio. They can be agglomerated to form particles,

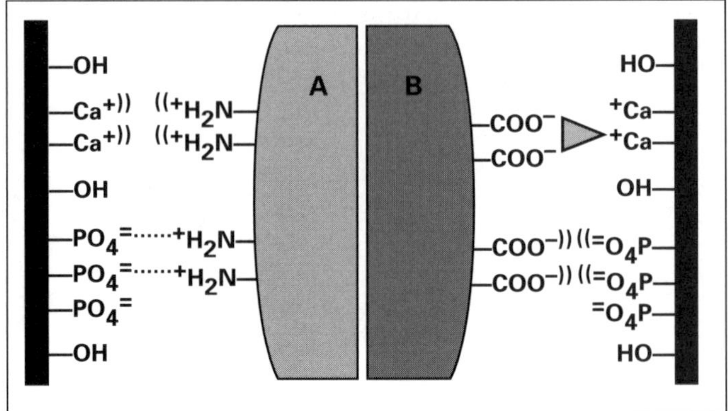

Figure 5.1. Protein binding to Hydroxyapatite. A is a basic protein. B is an acidic protein. Double parentheses indicate repulsion. Dotted lines indicate ionic bonds. Triangular linkages indicate coordination bonds.

heated to fuse the particles into a stable porous "ceramic" mass, and then sorted to produce populations of uniform particle size distribution. This yields media with flow properties, capacity, and scale-up attributes competitive with other popular methods.[9-12]

protein interactions

The interactions of proteins with hydroxyapatite are complex. The following discussion is a simplification but it includes the points most germane to antibody purification. Amino groups are attracted to P-sites but repelled by C-sites. The situation is reversed for carboxyls (Figure 5.1).[13-16] Although amine-binding to P-sites and the initial attraction of carboxyls to C-sites are electrostatic, the actual binding of carboxyls to C-sites involves formation of much stronger coordination complexes between C-sites and clusters of protein carboxyls. This has been proven experimentally by evaluating retention of proteins on which the carboxyls have been replaced by sulfo groups.[15] Binding is reduced dramatically even though net charge is unaltered.[15] Further proof that carboxyl/C-site binding does not reflect a classical anion exchange interaction is found in the fact that binding capacity diminishes for acidic proteins with increasing pH.[14,15,17,18]

phosphoryl interactions

Phosphoryl groups on proteins and other solutes interact even more strongly with C-sites than do carboxyls.[13] This is reflected in extremely strong binding by phosphoproteins.[19] DNA does not bind as strongly as expected for a phosphoryl-rich solute. The spacing of the phosphoryl groups along the backbone

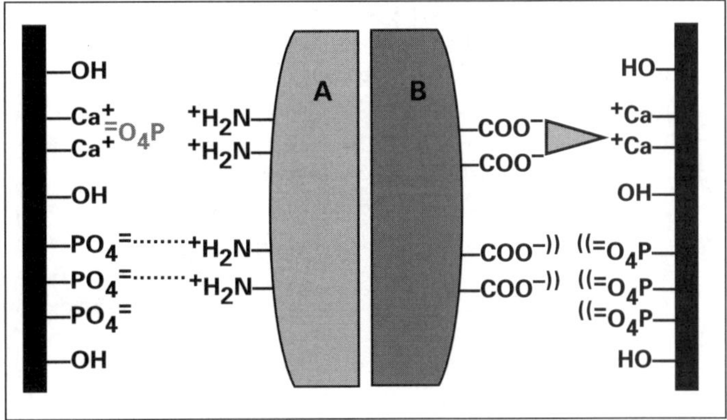

Figure 5.2. Enhanced amine-binding by blocking C-site repulsion. A is a basic protein. B is an acidic protein. Double parentheses indicate repulsion. Dotted lines indicate ionic bonds. Triangular linkages indicate coordination bonds. Ion pairing of buffer phosphates with C-sites suppresses amine repulsion. Compare with Figure 5.1.

apparently prevents an ideal match with the steric distribution of C-sites.[13,20-23] DNA binds well nonetheless and the strength of the interaction increases with its size. Endotoxins bind by the numerous phosphoryl groups on their core polysaccharide and lipid-A moieties.[24]

binding of basic proteins

Binding of basic proteins becomes stronger with reducing pH, due to increasing positive charge on the protein. This reflects the dominant cation exchange component of the interaction, but the selectivity is distinct from classical cation exchange. Concurrent repellence of amines by C-sites, and the geometric distribution of charges, impart a unique stereochemical element that sometimes endows HAC with the ability to discriminate among closely related protein variants. Examples include fractionation of light chain idiotypes from monoclonal mixtures with common heavy chains and fractionation of bifunctional antibodies from complex parent/sibling mixtures.[25-32]

Binding of weakly interacting basic proteins can be strengthened by inclusion of 1mM phosphate in the buffer.[14] The free phosphate ions pair with C-sites and suppress their ability to repel amines (Figure 5.2).This low concentration does not interfere with ionic binding between amines and P-sites. Elution of basic proteins can be accomplished with chloride or phosphate ions from 0.05–0.5M.[13-16] Although not reported, it should be possible to elute basic proteins with an increasing pH gradient. Descending pH gra-

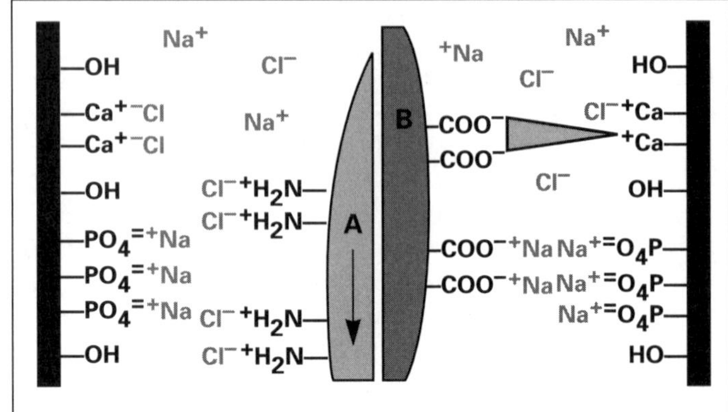

Figure 5.3. Selective dissociation of amine binding with sodium chloride. A is a basic protein. B is an acidic protein. Triangular linkages indicate coordination bonds. Note that coordination bonds are unaffected.

dients have been applied, exploiting titration of the P-sites, but this is risky. The crystal structure of hydroxyapatite degrades rapidly below pH 5.[13] Antibodies elute in the range of pH 2.5–4.5.[29] Even if the matrix were stable, the pH would still be a concern from the standpoint of product denaturation

selective elution of acidic solutes

Acidic solutes will not elute in sodium chloride, even at concentrations >3.0M (Figure 5.3). Elution requires displacers with stronger affinity for C-sites, such as phosphate, citrate, or fluoride ions. This has important ramifications for antibodies that behave as basic proteins. It means that they can be eluted with sodium chloride, completely avoiding risk of contamination from the bulk of acidic sample components.[13,33] This specifically includes the ability to elute IgG under conditions where DNA and endotoxin remain quantitatively bound.

For antibodies that bind dominantly as acidic proteins, this means that they can be applied to the column in a sufficient concentration of sodium chloride to maintain their solubility during loading.[25] Tolerance of high sodium chloride allows dissociation of ionic complexes between antibodies and acidic contaminants like DNA, thereby increasing purification performance and product binding capacity. High sodium chloride tolerance also allows sample to be loaded with no equilibration other than pH titration.

Attributes

HAC is applicable to all antibodies, providing good fractionation regardless of the production medium,

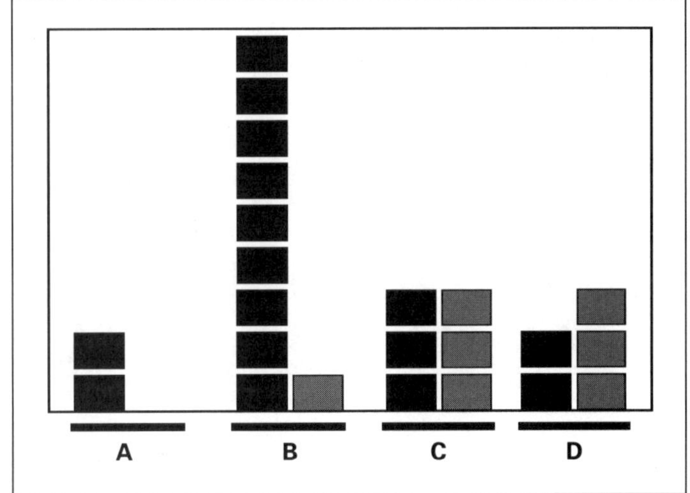

Figure 5.4. Frequency histogram of reported phosphate elution molarities for antibodies. Black units indicate IgG. Gray units indicate IgM. Most of these results were obtained at pH 6.8
A: 0.0–0.1M
B: 0.1–0.2M
C: 0.2–0.3M
D: 0.3–0.4M

species, class or subclass. Its purification capabilities are on a par with IEC, HIC, and immobilized metal affinity (IMAC), sometimes achieving better than 90% purity from raw production media in a single step. Frequent contaminants with phosphate elution include low levels of albumin, transferrin, and β_1-haptoglobin.[25-39] Contaminating polyclonal IgG can usually be reduced 50-75%, depending on the elution characteristics of the particular monoclonal.[27] Aggregates bind more strongly than native proteins.[25]

DNA clearance

DNA clearance in the traditional phosphate gradient mode is variable depending on the retention properties of the antibody and the size distribution of the DNA. Chromosomal DNA elutes in the range of 0.20–0.27M phosphate.[13] Fragments begin to elute at ~0.1M.[40,41] Figure 5.4 illustrates a frequency histogram of reported phosphate elution molarities for antibodies. In one study, phosphate-mediated HAC gave DNA clearance equivalent to anion exchange: modest at best, ~20 fold.[42] This emphasizes the merit of eluting antibodies with sodium chloride whenever possible. DNA clearance in that system should rank with—or outrank—the best alternatives.

viral clearance

HAC supports viral clearance of 0.6–3.5 logs.[42,43] In one study it provided less average clearance than other purification methods.[43] In another, it outper-

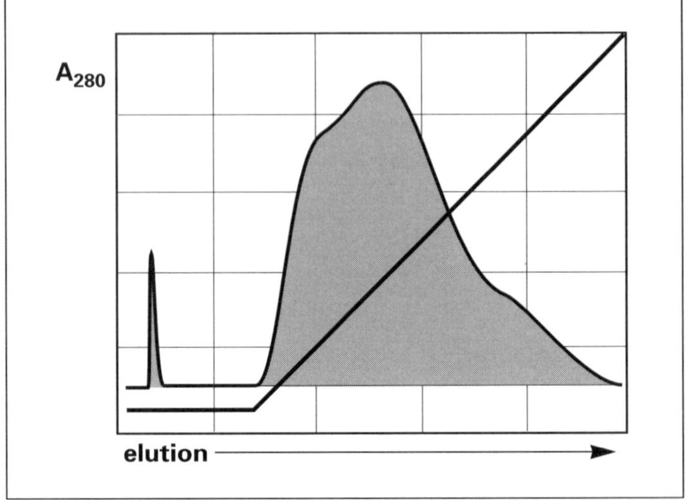

Figure 5.5. Endotoxin elution from hydroxyapatite. A commercial endotoxin preparation was injected onto a column equilibrated with 0.05M MES, pH 5.6, then eluted with a linear gradient from 0.0–1.0M potassium phosphate.

formed SEC by a factor of 15, IEC by a factor of 75, and salt precipitation by a factor >150.[42] Differential clearance abilities of the chloride and phosphate elution modes have not been characterized.

endotoxin clearance

Endotoxin removal can exceed 3 logs for antibodies that elute in sodium chloride. Clearance for antibodies that require phosphate elution is variable but seldom as good as 10-fold. As shown in Figure 5.5, endotoxin elutes continuously from 0.0–1.0M phosphate. As with DNA, clearance efficiency depends on the relative elution position of the antibody. Removal of phenol red is similarly antibody-dependent.[26]

mass recovery

Mass recoveries of 90% are typical. Recoveries below 80% usually reflect inadequate method development. Frequent causes include inadequate binding conditions, inadequate column equilibration, poor antibody solubility under the loading conditions, and excessively narrow pooling.

activity recovery

Activity recovery per mass unit of antibody is typically quantitative, but may be reduced with IgMs that are exposed to low ionic strength for prolonged periods awaiting chromatography.[25]

high throughput

Because of their high mechanical strength and open macroporous architecture, ceramic hydroxyapatite media support excellent capture efficiency at high flow rates. They easily support flow rates of 600cm/hr without significant loss of dynamic capacity or resolution. This enables rapid method development and processing at all scales.

ease of packing

Ceramic HAC media pack like sand; column packing quality is virtually never an issue. Ceramic HAC beds are also tolerant of introduced air, at least to the extent that it doesn't decimate packing quality as it does with soft gels. Introduced air can be removed with an upward flow of degassed solvent.

The price you pay for easily packed air-stable beds is that the high density of the media requires a creative approach to unpacking. Simply flow the column from the bottom. This will fluidize the bed, which can then be pumped out or decanted. Avoid tools that may fracture the media and create fines.[44]

resistance to harsh sanitizing conditions

With the exception of chelating agents and pH below 5—both of which degrade crystal structure—HAC media are resistant to the harshest cleaning agents, including concentrated sodium hydroxide, urea and guanidine, organic solvents, and detergents.[13-16] This turns out to be important because the worst foulants in chromatography are mostly phosphorylated and bind strongly to HAC media. Besides DNA and endotoxins, they include lipoproteins and free phospholipids. Large protein aggregates also bind strongly. Rather than rely on harsh cleaning methods, it's best to remove foulants in advance. A broadly applicable method is described in appendix II.

cleaning

Cleaning follows a different approach than is generally used for polymer based media. Begin with a 5CV linear gradient to 1.0M potassium phosphate. This will remove most strongly bound materials. Follow with 5CV of 0.1mM Triton X-100, 0.05M sodium hydroxide. Keep the solution on the column for at least 2 hours, maintaining a low flow rate to remove contaminants as they are solubilized. If you are using the media to purify monoclonal antibodies for in vivo application, you may wish to evaluate longer exposure for enhanced microbial kill and destruction of endotoxin. Apply a 5CV gradient to 60% methanol. Store the column in 60% methanol.

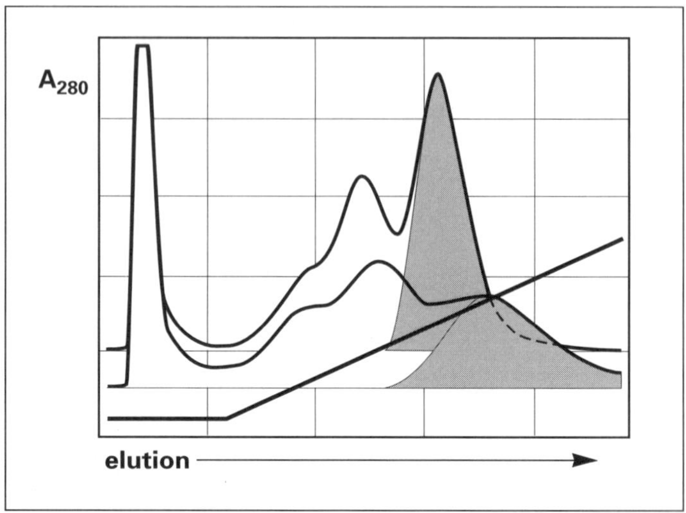

Figure 5.6. The effect of solubilizing sodium chloride in the sample buffer. Sample: Mouse IgM in ascites. The column equilibration and sample buffer for the upper profile was 5mM sodium phosphate, 0.1M sodium chloride, pH 6.8. Recovery was >90%. Corresponding buffers for the lower trace were the same except lacking sodium chloride. Recovery was ~75%.

Limitations

The worst of HAC's historical liabilities derive from a phenomenon that has gone almost unrecognized by suppliers and users alike: matrix degradation from acid secretion by microbial contaminants. This causes gross loss of separation performance, capacity, reproducibility, and column life in as little as 24 hours—even at 4°C.[25,45] The only preventative agent confirmed to date is a 50–60% aqueous solution of methanol.[25] Column performance was unchanged for a several-week period when it was used between runs as a preservative. Such a high concentration of a flammable reagent is undesirable to say the least. Less hazardous formulations will hopefully be developed, but demand rigorous validation of equivalence before you accept any alternative.

antibody solubility limitations

A second historical problem with HAC has been poor antibody solubility under sample loading conditions. The majority of published accounts recommend that antibodies be applied in 10mM phosphate buffer at pH 6.8. Virtually all IgMs and many IgGs are either partially or wholly insoluble under these conditions. The problem is not in evidence with analytical scale injections of unequilibrated sample on small columns because binding is achieved before aggregation or precipitation can occur. At preparative scale, samples equilibrated in advance have ample time to respond to the

Table 5.1. Dynamic capacity as a function of sintering level. Data for each particle size/sintering level were averaged from reference 48. BSA is bovine serum albumin. LYS is egg-white lysozyme.

Diameter µm	sintering level	capacity BSA	capacity LYS
20	A	2.1	14.9
20	B	25.4	21.4
40	A	4.3	14.9
40	B	24.3	22.0
80	A	6.2	14.3
80	B	21.9	21.4

conditions. This causes peak broadening. Aggregates often elute in a separate peak—or smear. Mass recovery can be severely depressed (Figure 5.6).

There are simple remedies. Most antibodies tolerate significant concentrations of sodium chloride without loss of binding efficiency. Inclusion of 0.1M sodium chloride in the loading buffer has been observed to suspend all solubility problems with IgMs.[25,46,47] The other solution, for antibodies that won't tolerate elevated conductivity, is to apply sample by on-line dilution, as is done with cation exchange chromatography.

media issues

The third historical liability, and the one most widely recognized, has been the unavailability of HAC media with the physical strength and flow properties to support commercial use. Ceramic media have resolved these limitations but there are important variations to be aware of. Differences in the sintering (heat fusion) process alter the relative accessibility of P-sites and C-sites.[13,48] This is reflected in the comparative binding capacities for basic and acidic proteins (Table 5.1).

Seek media with the most "native" hydroxyapatite characteristics, as indicated by similar capacities for acidic and basic proteins. Manufacturers commonly express these values as capacities for bovine serum albumin and lysozyme. Watch for variations in selectivity with different particle sizes. They can affect scale-up.

glycoform fractionation

As with IEC, fractionation of glycosylation isoforms is important to watch for. Systematic comparisons have not been made, but the frequency of occurrence seems

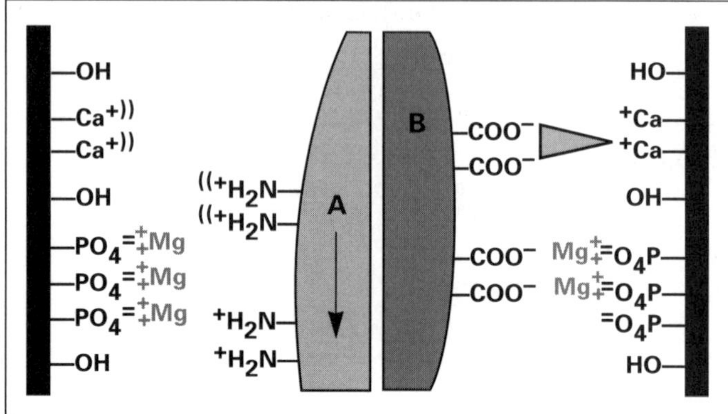

Figure 5.7. The effect of divalent metal cations on selectivity. A is a basic protein. B is an acidic protein. Double parentheses indicate repulsion. Triangular linkages indicate coordination bonds. Charge pairing of divalent metal cations with P-sites blocks amine binding at the same time it suppresses repulsion of negatively charged groups. This causes stronger binding of acidic solutes.

to be higher with HAC than with IEC. This is a concern because glycosylation affects antibody titer, pharmacokinetics, and stability.[49-55] This does not mean that all changes in glycoform distribution have dire consequences for antibody performance. It does mean that you should use IEF to evaluate pooling specifications, avoid specifications that alter glycoform distribution, and validate any distributional shifts that do occur. If glycoform fractionation is your intent, 5µm media will maximize separation quality.

antibody complexation with DNA

HAC is vulnerable to the same idiosyncrasies of monoclonal behavior that afflict other separation methods. One of these is electrostatic complexation of DNA with strongly basic IgGs and IgMs. These antibodies can transport DNA through a process despite the selectivity of the media. The problem is compounded by loss of separation performance and depressed product recovery.[56-58] Sodium chloride concentrations from 0.3–1.0M will dissociate the complexes for antibodies with sufficiently tolerant HAC binding properties, but the only recourse for others is advance DNA removal. Both HIC and IMAC fulfill this need (chapters 6 and 7).

metal contamination

Metal contamination of process buffers is a concern because complexation with P-sites modifies matrix selectivity.[13-16] Amine binding can be abolished entirely (Figure 5.7). Achieving this level of selectivity modification requires deliberate metal-loading of the column but it illustrates the type of aberration that metal contamination can cause. Metals also bind to

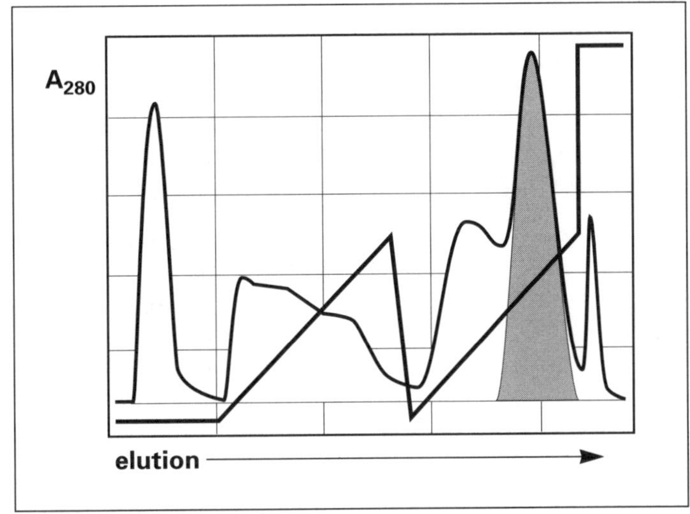

Figure 5.8. Dual gradient screening. Mouse IgG_1 ascites. The first gradient goes from 0.0–0.5M sodium chloride. The second goes from 0.0–0.5M potassium phosphate. See Table 5.2 for buffers and conditions. The majority of IgGs, virtually all IgMs, and IgAs behave much as shown.

proteins, altering their charge characteristics.[59] Even if you can tolerate the shift in selectivity, they destabilize proteins.[60,61] Chelating agents are not a control option because they dissolve the media.[13] The only course is to use high quality salts.

temperature considerations

HAC separations are usually run at room temperture. If you plan to conduct cold processing, you need to be wary of cryoglobulins. Cryoprecipitation is rarely a problem with IgGs but occurs frequently with IgMs.[62-66] Inclusion of 1.0M urea in all buffers is usually the most effective suppressant. At this concentration, it neither interferes with binding nor causes serious concerns about denaturation.[67-69]

buffer incompatibilities

A minority of strongly basic antibodies are able to form stable ionic crosslinks with polyvalent anions such as phosphate. This is seen most often among IgG_3s, though not exclusively. The consequent aggregation severely compromises chromatographic purification and recovery, regardless of the separation mechanism. If these antibodies won't elute in sodium chloride, use sodium fluoride.

Method development

Method development for HAC follows the same general format as IEC (see Chapter 4). Only points of difference are discussed here. The first step is to determine whether the antibody behaves as dominantly basic or dominantly acidic. This can be done in a single

Table 5.2. Buffers and conditions for dual gradient screening. The low pH maximizes binding efficiency and capacity but is still a full pH unit above the point where the matrix becomes unstable.

Buffers

A: 0.05M MES, 1mM potassium phosphate, pH 6.0
B: A + 0.50M sodium chloride
C: A + 1.0M potassium phosphate

Conditions

Equilibrate column: with 10CV buffer A
Inject: 2%CV unequilibrated sample
Wash: 2CV buffer A
Elute: in a 10CV linear gradient to buffer B
Wash: with a 2CV linear gradient to buffer A
Elute: in a 10CV linear gradient to 50% buffer C
Strip: with 5CV buffer C

run by eluting the column with sequential gradients; first with sodium chloride, then with potassium phosphate (Figure 5.8).[13] See Table 5.2 for conditions.

If the antibody elutes at a relatively high sodium chloride concentration, rescreen without phosphate. Amine-repulsion by C-sites will prevent binding by a large subset of basic contaminants. If the antibody still binds strongly, substitute 0.05M HEPES at pH 7.0. If it still binds strongly, try BICINE at pH 8.0. Every rung you climb the ladder will reduce the number of contaminants coeluting with the antibody.

If the antibody elutes only in phosphate, run a series of screens, loading the column with increments of sodium chloride up to 1.0M. This will clear basic contaminants from the system as well as dissociate ionic complexes among contaminants. It will also identify antibodies that can be loaded without dilution. You can omit the sodium chloride after the sample load and wash. It has no effect on selectivity of the phosphate elution and provides no other benefits.

choice of elution modes

If an antibody elutes in sodium chloride, this is the best preparative mode to pursue. It will provide the best protein purity in addition to the best DNA and endotoxin clearance. Once you've selected the binding and elution modes, prepare enough material so that you can compare purification performance with other candidate methods. (See chapter 4).

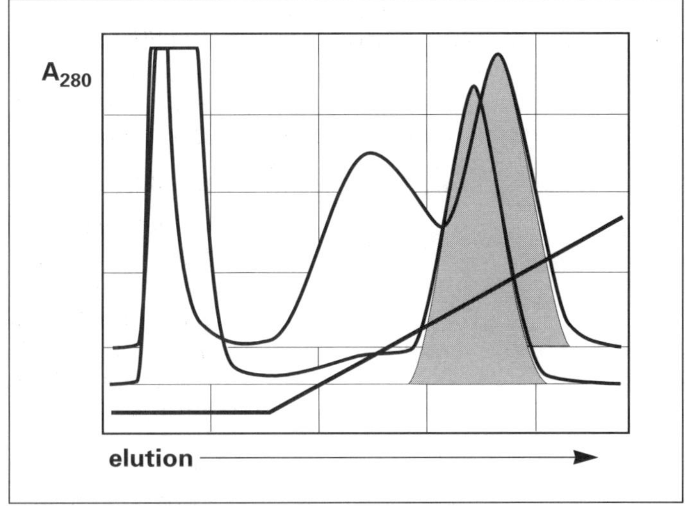

Figure 5.9. The effect of phosphate concentration on protein binding. The upper profile was loaded with 1mM phosphate in the binding buffer. The binding buffer for the lower profile included 0.05M phosphate. IgG binding capacity was improved by reduced substrate competition with contaminants.

process order

Placement of HAC within a multistep process depends on an antibody's elution behavior. If it elutes in sodium chloride, most contaminants will bind more strongly than the antibody and tie up column capacity. This doesn't mean that you can't use HAC for a first step, just that you'll have to compensate with a larger bed volume. If binding tolerates a high level of sodium chloride, the convenience of loading sample without dilution or buffer exchange may justify the inflated bed volume. If an antibody elutes at a high phosphate concentration, conditions can be set that prevent binding by most contaminants (Figure 5.9).

scale-up issues

Most scale-up and manufacturing problems with HAC derive from inadequate column maintenance or inappropriate binding conditions. Make sure that cleaning and storage protocols are tailored to the media's special needs, and make sure that the antibody is fully soluble under loading conditions. Otherwise, scale-up problems occur for the same reasons as with other adsorptive methods (see chapter 4).

In perspective

HAC has not been exploited to its potential as a process tool for monoclonal purification, mainly for lack of suitable media and monoclonal-specific application guidelines. With proper maintenance and method development customized to its unique characteristics, HAC offers process capabilities on a par with the best alternatives. Recent developments in ceramic media provide all the performance characteristics needed for effective scale-up.

purification of in vitro products

HAC supports a variety of 2-step procedures for purification of in vitro products. HAC combined with SEC is effective for IgMs; infrequently so for IgGs. Two-step combinations with HIC can be applied to both classes, generally supporting better purification, higher capacities, and better overall process economy. Combination with IMAC is worth evaluation. Combinations with IEC may provide adequate performance, but combinations of charge-based methods are generally less effective than partnerships between dissimilar separation mechanisms.

purification of in vivo products

For in vivo applications, HAC may provide a useful selectivity alternative to IEC. Its nucleic acid clearance capabilities take on special value for antibodies that elute in chloride. Processes that combine either high-salt IMAC or HIC, with chloride-HAC, especially in that order, although not necessarily in immediate sequence, promise better DNA and endotoxin removal than any other combination.

Recommended reading

The article by Josic, addressing matrix degradation by microbial contamination, is essential reading for anyone using HAC.[25] For discussion of retention mechanisms, begin with the series by Gorbunoff and continue with the review by Kawasaki.[13-16]

References

1. A. Tiselius et al, 1956, *Arch. Biochem. Biophys.*, **65** 132
2. T. Kawasaki et al, 1985, *J. Biochem.*, **152** 361
3. T. Kawasaki, 1978, *J. Chromatogr.*, **151** 95
4. T. Kawasaki, 1978, *J. Chromatogr.*, **157** 7
5. A. Atkinson, 1973, *J. Appl. Chem. Biotechnol.*, **13** 167
6. G. Bernardi,1971, *Met. Enzymol.*, **22** 325
7. M. Spencer, 1978, *J. Chromatogr.*, **166** 423
8. M. Spencer, 1978, *J. Chromatogr.*, **166** 435
9. T. Kadoya, 1986, *J. Liquid Chromatogr.*, **9** 3543
10. T. Kawasaki et al, 1986, *Eur. J. Biochem.*, **157** 291
11. T. Kawasaki et al, 1986, *Biochem. Int.*, **13** 969
12. Y. Kato et al, 1988, *J. Chromatogr.*, **398** 340
13. T. Kawasaki, 1991, *J. Chromatogr.*, **544** 147
14. M. Gorbunoff, 1984, *Anal. Biochem.*, **136** 425
15. M. Gorbunoff, 1984, *Anal. Biochem.*, **136** 433
16. M. Gorbunoff and S. Timasheff, 1984, *Anal. Biochem.*, **136** 440

17. G. Bernardi et al, 1972, Biochim. Biophys. Acta, 160 301
18. T. Ogawa and T. Hiraide, 1995, Effect of pH on Gradient Elution of Proteins on Two Types of Macroprep Ceramic Hydroxyapatite, poster, Prep-Tech '95, E. Rutherford, NJ
19. G. Bernardi and W. Cook, 1960, *Biochim. Biophys. Acta*, **44** 96
20. H. Martinson, 1973, *Biochemistry*, **12** 2731
21. H. Martinson and E. Wagenaar, 1974, *Biochemistry*, **13** 1641
22. G. Bernardi et al, 1972, *Biochim. Biophys. Acta*, **278** 409
23. G. Bernardi, 1971, *Met. Enzymol.*, **21** 95
24. J. Homma (ed.), 1984, Bacterial Endotoxins: Chemical, Biological, and Clinical Aspects, Verlag Chemie, Basel
25. Dj. Josic et al, 1991, *Hoppe Seylar's Z. Physiol. Chem.*, **372** 149
26. J. Bukovsky and R. Kennett, 1987, *Hybridoma*, **6** 219
27. T. Brooks and A. Stevens, 1985, *Am. Lab.*, **Oct**. 54
28. H. Juarez-Salinas et al, 1984, *BioTechniques*, **2** 164
29. L. Stanker et al, 1985, *J. Immunol. Met.*, **76** 157
30. H. Juarez-Salinas et al, 1986, *Met. Enzymol.*, **121** 615
31. M. Mariani et al, 1989, *Immun. Today*, **10** 115
32. M. Mariani et al, 1989, *BioChromatography*, **4** 149
33. T. Kawasaki et al, *Eur. J. Biochem.*, **155** 249
34. R. Sim and R. Discipio, 1982, *Biochem. J.*, **205** 285
35. Dj. Josic et al, 1988, *J. Clin. Chem. Clin. Biochem.*, **26** 559
36. G. Smith et al, 1984, *Anal. Biochem.*, **141** 432
37. C. Poesi et al, 1989, *J. Chromatogr.*, **465** 101
38. L. Manil et al, 1986, *J. Immunol. Met.*, **90** 25
39. Y. Yamakawa and J. Chiba, 1988, *J. Liquid Chromatogr.*, **11** 665
40. HCA Column, techical information 4 (TD-505), Mitsui Toatsu Chemical Company, Tokyo
41. HCA Column, techical information 6 (TD-507), Mitsui Toatsu Chemical Company, Tokyo
42. G. Dove et al, 1990, ACS Symposium Series #427, p.194, American Chemical Society, Washington D.C.
43. J. Grun et al, 1992, *BioPharm*, **5**(9) 22
44. A. Moschella, 1996, personal communication, Bioprocessing Inc., Princeton, NJ USA
45. P. Steffens, 1989, *GIT Fachz. Lab. Suppl. (Chromatogr.)*, **389** 50
46. J. Porath, 1964, *Biochim. Biophys. Acta*, **90** 324
47. U.-B. Hansson and E. Nilsson, 1973, *J. Immunol. Met.*, **2** 221
48. L. Cummings et al, 1995, Macroprep Ceramic Hydroxyapatite—New Life for an Old Chromatographic technique, Bulletin 1927 US/EG Rev. A (95-0143 0295) Bio-Rad Laboratories, Hercules, CA USA
49. R. Parehk, 1992, *Biotech. Lab.*, **11** 61

50. Y. Kagawa, 1988, *J. Biol. Chem.*, **263** 508
51. M. Malaise, 1987, *Clin. Immunol., Immunopathol.*, **4** 51
52. B. Maiorella et al, 1993, *Bio/Technology*, **11**(3) 387
53. T. Monica et al, 1993, *Bio/Technology*, **11**(4) 512
54. C. Goochee et al, 1990, *Bio/Technology*, **8**(5) 421
55. C. Goochee et al, 1992, Frontiers in Bioprocessing II, (P. Todd et al, eds.) p.199, American Chemical Society, Washington, D.C.
56. P. Gagnon, 1996, Special Weapons and Tactics for Removal of Product-Bound DNA, slide presentation, BioEast '96, Washington D. C.
57. R. Steiner, 1953, *Arch. Biochem. Biophys.*, **46** 291
58. R. Steiner, 1953, *Arch. Biochem. Biophys.*, **47** 56
59. F. Gurd and P. Wilcox, 1956, *Adv. Protein Chem.*, **11** 312
60. T. Arakawa and S. Timasheff, 1982, *Biochemistry*, **21** 6545
61. T. Arakawa and S. Timasheff, 1984, *Biochemistry*, **23** 5912
62. C. Middaugh and G. Litman, 1977, *J. Biol. Chem.*, **252** 8002
63. G. Litman et al, 1981, *Immunol. Commun.*, **10** 707
64. C. Middaugh et al, 1981, *J. Biol. Chem.*, **255** 6532
65. K. Weber and L. Clem, 1981, *J. Immunol.*, **127** 300
66. C. Middaugh et al, 1978, *Proc. Nat. Acad. Sci. USA*, **75** 3440
67. C. Tanford, 1968, *Adv. Protein Chem.*, **23** 121
68. S. Timasheff and G. Fasman, 1969, Structure and Stability of Biological Macromolecules, Marcel Dekker, New York
69. W. Jencks, 1969, Catalysis in Chemistry and Enzymology, McGraw-Hill, New York

Chapter 6

Hydrophobic Interaction Chromatography

"You can't expect to hit the jackpot if you don't put a few nickels in the machine."
—Flip Wilson

The foundations of hydrophobic interaction chromatography (HIC) are the foundations of adsorption chromatography itself, with hydrophobicity having first been exploited as an adsorption mechanism in the late 1940s.[1] HIC has since evolved into one of the most powerful techniques in preparative biochemistry. Antibodies have been one of the chief beneficiaries. Their strong hydrophobicity, compared to the bulk of their contaminants, makes them ideal candidates for this technique.

Mechanisms

Protein retention behavior in HIC reflects cooperativity between 2 distinct mechanisms. The first is preferential exclusion of precipitating salts from protein and chromatography media surfaces. This leaves both preferentially hydrated.[2-5] With increasing salt concentration it becomes energetically favorable for proteins to share their hydration shells with the media, and they are retained.

This in itself is not HIC. Cohydration/cosolvent exclusion promotes chromatographic retention even in the absence of hydrophobic interactions. This is illustrated by protein retention in high salt solutions on size exclusion (SEC) media (Figure 6.1).[6,7] Analogous high-salt enhancement of binding on immobilized metal affinity (IMAC) columns has been shown specifically not to involve hydrophobic interactions.[8] The ability of proteins to bind to ion exchange columns

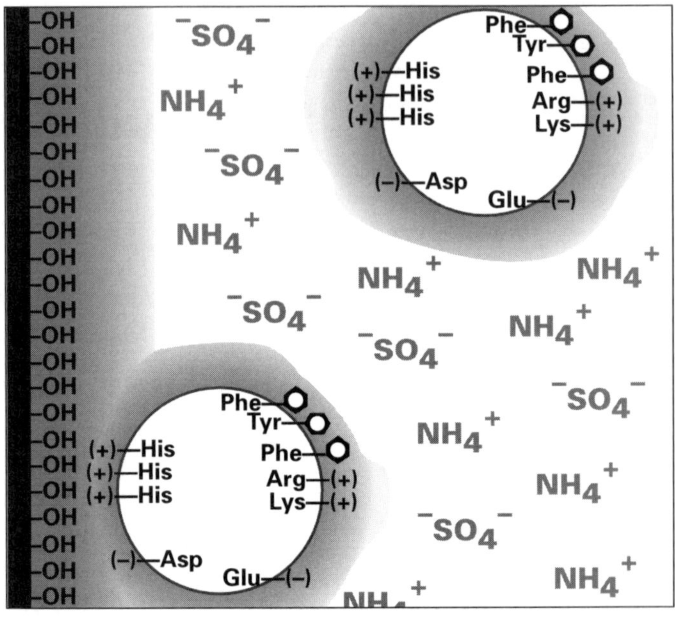

Figure 6.1. Protein adsorption to a nonhydrophobic support by cohydration/cosovent exclusion. See text for discussion.

(IEC) at salt concentrations many times higher than their normal elution levels likewise distinguishes salt-driven cohydration from HIC.

The defining characteristic of HIC is that the cohydrational associations are augmented and stabilized by hydrophobic interactions (Figure 6.2). Hydrophobic interactions are a strong attractive force in salt solutions.[9,10] They exert up to 2 orders of magnitude greater influence than van der Waals forces at distances up to 8nm. The strength of these interactions is proportional to the hydrophobicities of the attractive surfaces.[9] Hence strongly hydrophobic proteins bind more avidly than weakly hydrophobic proteins, and binding is stronger on media with stronger ligands and/or higher ligand densities.[10-16]

This dual mechanism has important performance ramifications for antibody purification. Weakly hydrophobic media provide little enhancement to cohydrational adsorption. This means that they require a very high level of precipitating salt to promote adsorption, but also that antibodies can be eluted quantitatively simply by restoring solvation water, i.e., removing the salt. Strongly hydrophobic media support equivalent

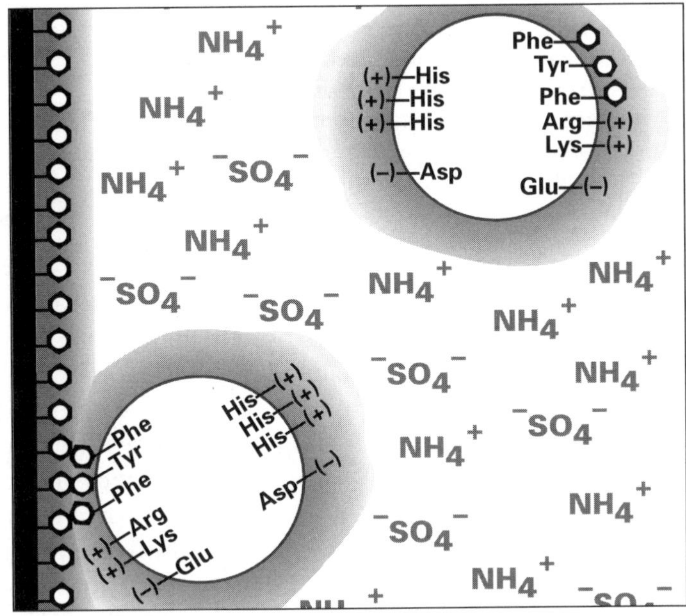

Figure 6.1. Protein adsorption by a combination of cohydration/ co-sovent exclusion with hydrophobic interactions. See text for discussion.

adsorption efficiency at lower salt concentrations because the stronger hydrophobic interactions overcompensate for diminished cohydration. However, suspension of cohydration is not adequate to achieve quantitative elution. Even in the absence of precipitating salts, elution from these media requires direct competition with the hydrophobic component of the mechanism, for example by addition of organic solvents. This accounts for the relatively broad peaks and reduced recoveries encountered when strongly hydrophobic media are eluted solely by reducing salt gradients (Figure 6.3).

effects of different salts

The parallel between the ability of various salts to promote adsorption and the rankings of their component ions in the Hofmeister series of lyotropic ions is well documented (Table 6.1).[17-33] The differences are mediated chiefly through the cohydration component of the mechanism. Salts composed of ions high up in the lyotropic series are strongly excluded from protein and media surfaces.[2-5] Salts composed of ions from the chaotropic region of the series penetrate hydration shells, binding directly to protein and media surfaces. This reduces local hydration; cohydrative

Table 6.1. The Hofmeister series of lyotropic and chaotropic ions. The lyotropic effect is sometimes referred to as the "salting-out" effect, while the chaotropic effect is referred to as "salting in".

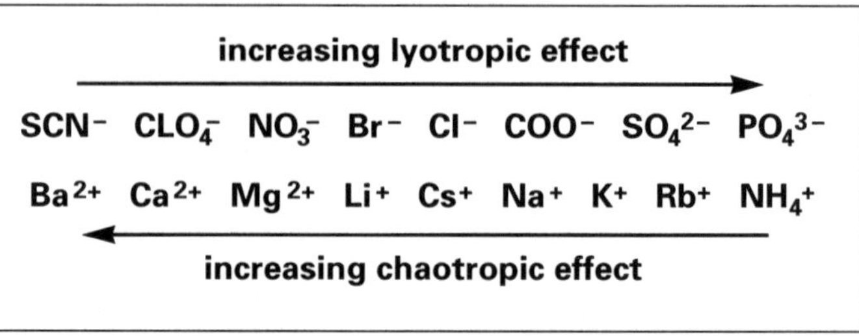

increasing lyotropic effect →								
SCN^-	CLO_4^-	NO_3^-	Br^-	Cl^-	COO^-	SO_4^{2-}	PO_4^{3-}	
Ba^{2+}	Ca^{2+}	Mg^{2+}	Li^+	Cs^+	Na^+	K^+	Rb^+	NH_4^+
← increasing chaotropic effect								

adsorption with it. Divalent metal cations in particular are very strongly bound and strongly dehydrative.

choice of bindng salt

Variation in the balance of surface exclusion/binding properties among different salts is a primary source of selectivity differences in HIC. This implies that choice of salts merits investigation as a routine method development parameter. However, ions that bind proteins destabilize proteins.[2,3] It also requires excessive salt concentrations to overcome their dehydrative effects. This suggests that binding salts should be limited to those composed of ions high up in the lyotropic series.[34] This conclusion is supported by performance-based studies showing that the protein-stabilizing effects of such salts translate into sharper peaks and better activity recovery.[20,24,28,35,36]

Ammonium sulfate has historically been the HIC salt of choice, but it bears an important limitation: its preparative application must be restricted to neutral and acidic pH. Liberation of free ammonia at alkaline pH causes protein denaturation.[24,37,38] Volatilization of the gaseous ammonia can be a health hazard, especially at large process scale.[34] It is also a process control problem. It destabilizes pH and partially subverts the ionic equilibrium of the dissolved salt from ammonium sulfate to sodium sulfate.[24,37-39]

Potassium phosphate provides an alternative for alkaline applications. Its average molar selectivity is similar to ammonium sulfate, although individual proteins respond differently (Figure 6.4).[34] Monosodium glutamate and other organic salts are also an option.[39] Like inorganic salts high in the lyotropic series, they are strongly excluded from protein surfaces, and hence protein-stabilizing.[40,41]

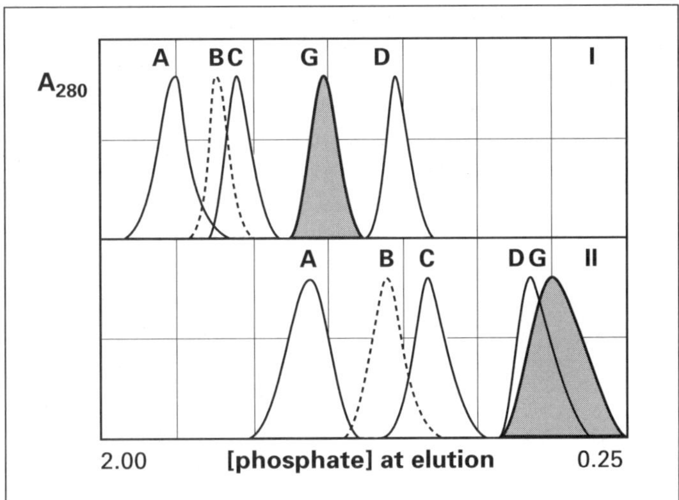

Figure 6.3. Variation in peak width as a function of column hydrophobicity. Series I was developed on weakly hydrophobic column; series II on a phenyl column. Column dimensions, base matrix, samples, buffers, and conditions were otherwise identical. A: lysozyme. B: chymotrypsinogen.C: transferrin. D: R-phycoerythrin. G: mouse IgG_1. Note that RPE retention is altered less by the shift in matrix hydrophobicity than the other proteins. This indicates that it's highly hydrated, but has relatively low surface hydrophobicity.

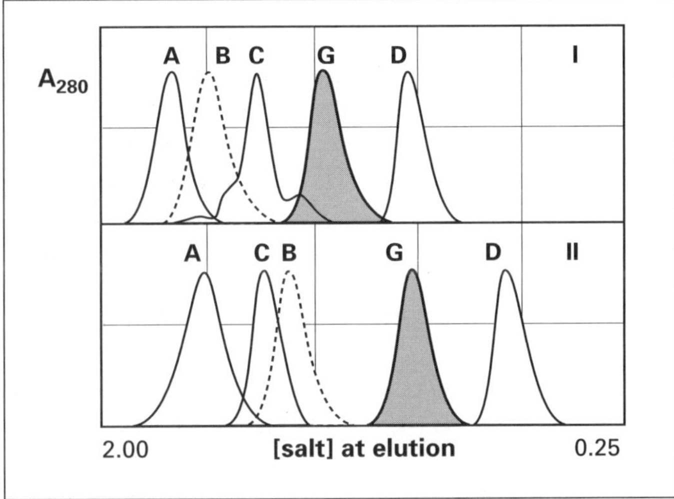

Figure 6.4. Variation in selectivity between binding salts. Both series were developed on a weakly hydrophobic column. Series I was developed with ammonium sulfate; series II with potassium phosphate. Column dimensions, base matrix, samples, and conditions were otherwise identical. A: lysozyme. B: transferrin. C: chymotrypsinogen. D: R-phycoerythrin. G: mouse IgG_1. The difference in peak configuration for chymotrysinogen reflects its interaction with either sulfate or ammonium ions.

The solubility of sodium sulfate and sodium phosphate are too limited to be useful for antibody purification on weakly hydrophobic media. Despite its >5.0M solubility, sodium chloride is too weak a promoter to support efficient adsorption.[34]

selectivity modifiers

Selectivity modifying additives can be divided into 2 groups; those that enhance binding by increasing cohydration efficiency, and those that weaken it by acting on the hydrophobic component of the mechanism. Enhancers include sucrose and glycine but neither appear to offer worthwhile selectivity modification.[34]

Glycine may nevertheless be of interest for its antibody solubilizing and stabilizing properties.[42-44] Included in HIC eluting buffers, it appears to improve recovery, especially of IgMs.

Hydrophobic competitors should be evaluated with extreme caution since most of them are strong denaturants. This is especially true of simple alcohols, detergents, and high concentrations of urea.[2,3,34,45-54] Ethylene glycol is a valuable exception. Like precipitating salts, it is excluded from protein surfaces at concentrations up to 30–60%, making it protein-stabilizing below these limits.[51,54] Above this range, it involves progressive risk of denaturation. An important caution with ethylene glycol: it absorbs up to twice its weight in water.[55] Good process control demands that it be qualified as a "dry" reagent and that storage conditions be validated to maintain it in this condition.[34]

pH effects

The effects of pH on selectivity are unpredictable despite the mechanism being well understood.[20,23, 34,37,56-62] When amphoteric protein residues are fully titrated, they are minimally hydrated and minimally influential on adjacent hydrophobic sites. When they are fully charged, strongly electrostricted hydration water depresses the hydrophobicity of adjacent nonpolar sites. The sign of the charge is inconsequential.[34,36,37,58]

preferential orientation

Studies of 3-dimensionally characterized proteins demonstrate the importance of hydrophobic and charged-site distribution on protein surfaces. Proteins orient themselves to present their most hydrophobic surface to the column. The characteristics of areas outside the interface have negligible influence.[36,58] This reinforces the rationale for net charge having no predictive value regarding retention. It also explains deviations of HIC selectivity from salt precipitation.

hydrophobic sites on IgG

Three-dimensional x-ray crystallographic analyses indicate 18 hydrophobic sites on IgG (Table 6.2, Figure 6.5).[63-65] These sites collectively represent an area greater than 3000 Å^2. The Fc region contains 10 of the 18; 3 each on the Cγ2 domains, 1 each at the junctures of the Cγ2 and Cγ3 domains, and a shared

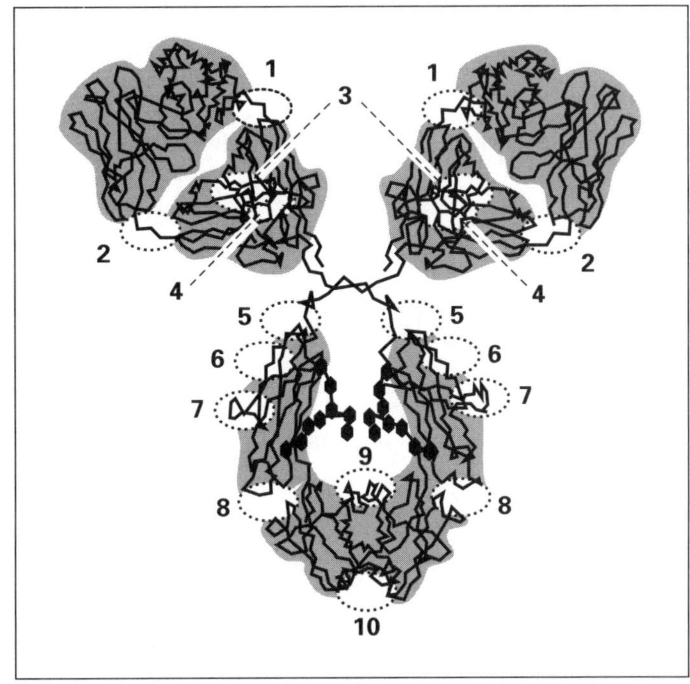

Figure 6.5. Hydrophobic sites on Human IgG_1. Black lines indicate the α-carbon skeleton as revealed by x-ray crystallography. Black hexagons indicate carbohydrate moieties. Dashed ovals indicate approximate location of hydrophobic sites. See Table 6.2 for legend to sites. Coordinated and redrawn from references 63-65.

Table 6.2. Composition of hydrophobic sites on Human IgG_1. The composition of these sites is highly conserved. Substitutions tend to be of character similar to the residues they replace. Charged residues with pKs between pH 4.5 and 8.5 are bolded. Refer to chapter 4, Table 4.1, for more information on pKs of charged residues in proteins.

Site	Principle residues	Area, Å^2
1. CL (a,b)	Pro, **Lys**, Ala (2),Thr (2)	320–390
2. CL (a,b)	Pro, Ala (3), **Lys**	190–230
3. Cγ1 (a,b)	**His**, Pro, Ala, Thr (3), Gly (3), Leu	400–500
4. Cγ1 (a,b)	Val, Leu (2), Ala, **Lys**, Gly	180–300
5. Cγ2 (a,b)	Ile, Leu, Met, **Gln**, **His**, Ser	400
6. Cγ2 (a,b)	Pro (2), **Lys (2)**, Ala (2), **Gln**	400
7. Cγ2 (a,b)	**Lys**, Pro, Arg (2), Ala, Val, **Glu**	280
8. Cγ2 (a,b)	Val (2), **His, Gln**	230
9. Cγ3	—	—
10. Cγ3	—	—

pair at the juncture of the Cγ3 domains. Each of the Fab regions contains four hydrophobic sites, arranged in symmetrical pairs on opposing axes. Which sites dominate binding to HIC media, and how consistently from monoclonal to monoclonal, is an intriguing but unanswered question.

hydrophobic site composition

All of the major hydrophobic sites are associated with at least 1 charged residue; some include up to 4.[63-65] The pK of the arginyl residues on the C-proximal bulge of the Cγ2 domains is ≥12, but the remainder are potentially titratable within the range of pH conditions normally exploited for HIC.[66-69] This is of interest because pH-dependent hydration of a single histidyl residue has been shown to have a significant effect on selectivity.[36,58] The presence of so many titratable residues within the hydrophobic sites of IgG emphasize the merit of exploring pH as a routine part of method development.[36,58]

Attributes

HIC is applicable to all monoclonals: all species; all classes; all subclasses; all production media. Single-step purity ranges from 60–90%, averaging ~80%. It easily removes the contaminants that are most difficult for SEC, IEC, IMAC and hydroxyapatite (HAC). Albumin is removed quantitatively regardless of its polymeric composition, except on excessively hydrophobic supports like octyl. Transferrin is removed regardless of its iron saturation state.

purification performance

HIC supports 50-75% removal of nonspecific antibodies. In combination with 1 IEC step, clearance often exceeds 90%; with both, it may exceed 99%. Antibody aggregates bind more strongly than native proteins. In some cases they are resolved into a separate peak. More typically they elute in a trailing shoulder.

DNA removal

HIC supports more efficient DNA removal than IEC, HAC, or SEC. Expect 3–5 logs of clearance—more with longer washes (Figure 6.6).[70-73] DNA is unretained by HIC media and the high salt concentration dissociates otherwise stable ionic complexes that transport it with the product through low-salt purification schemes.[70-76] See chapters 3 and 4 for more detail on antibody/DNA complexes.

virus removal

HIC supports 2–4 logs of viral clearance.[77-79] Its use for purification of some viruses suggests that the clearance mechanism relies in part on adsorption of viral lipid envelopes. This provides an additional dimension to multi-step viral reduction strategies.

endotoxin removal

HIC removes 1–2 logs of endotoxin.[71,72] Due to

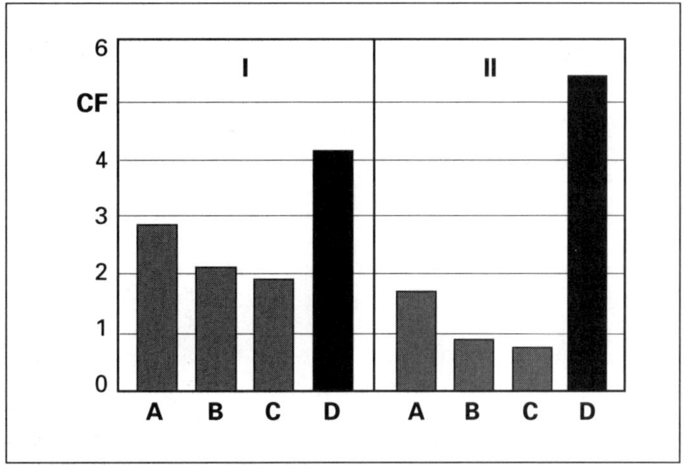

Figure 6.6. DNA clearance ability of HIC compared with other methods. Clearance factors (CF) given in $\log_{10}$. Series I illustrates clearance of total DNA from contaminated IgM cell culture supernatant. Series II illustrates clearance of DNA that was ionically complexed to the IgM in the supernatant. The sample for series II was from a cation exchange first-step. A: anion exchange. B: cation exchange. C: SEC. D: HIC

their detergent-like composition, endotoxins form large secondary structures in aqueous solution: micelles, ribbons, sheets, and vesicles with their polar residues external.[80] These structures are stabilized at high salt concentrations and are excluded from the matrix—passing through the void volume.

mass recovery

Specific mass recovery from moderately hydrophobic media is ~90%. Casual estimates are often lower because they fail to account for removal of contaminating antibodies. With low producing cell cultures that are heavily supplemented with serum, HIC can result in a 50% loss of total antibody while delivering 90% of the specific monoclonal. This distinction is important when comparing recoveries with nondiscriminating methods such as IMAC, SEC, protein A, or protein G. Specific recoveries from strong HIC media may be as low as 50%. This problem can be corrected to an extent by inclusion of up to 50% ethylene glycol in the eluting buffer. If recovery is not improved to at least 80%, use a less hydrophobic ligand.

activity recovery

Activity recovery is usually quantitative with weak-to-intermediate HIC supports. Apparent activity recoveries >100% (per mg of antibody) result from neglecting to adjust for removal of nonspecific antibody. Verified loss of specific immunoreactivity on an excessively hydrophobic support demands conversion to a weaker ligand.[81]

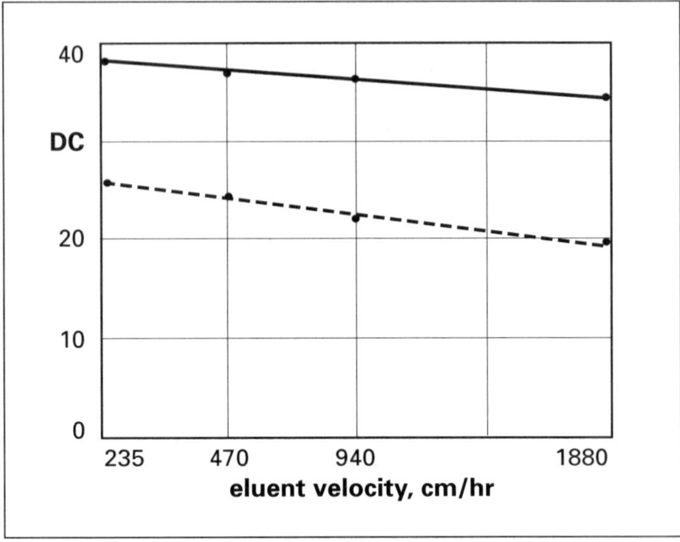

Figure 6.7. HIC dynamic capacity as a function of flow rate for 2 different mouse IgG_1 monoclonals under the same conditions. Measurements were made on a moderately hydryophobic column at 1.5M ammonium sulfate. DC signifies dynamic capacity. All values are indicated in mg/mL of gel.

high capacity

Most HIC media have dynamic capacities of at least 10mg/mL for IgG antibodies, and many exhibit double that or more.[82] As with all adsorption methods, dynamic capacity is strongly affected by a range of factors, particularly including composition of the mobile phase, eluent velocity, and contaminant competition (Figures 6.7 and 6.8).[82] Dynamic capacity under actual application conditions must be determined experimentally on a case-by-case basis (chapter 4).

high throughput

HIC fractionation can be conducted very rapidly. Most polymeric media support eluent velocities of at least 200cm/hr, allowing analytical runs to be conducted in ~30 minutes and preparative runs in 2–4 hours. The recent generation of media support analytical run times less than 5 minutes. Preparative runs can usually be completed in less than 30 minutes. This greatly encourages thorough process characterization, with large dividends during scale-up and validation.

resistance to harsh sanitizing conditions

Most HIC media withstand exposure to strong acids, bases, chaotropes, detergents, and organic solvents. This provides the basis for a virtually unlimited range of column regeneration and sanitization methods. Expect to obtain several hundred runs from prepacked HPLC columns. The life of preparative columns varies according to the composition of the

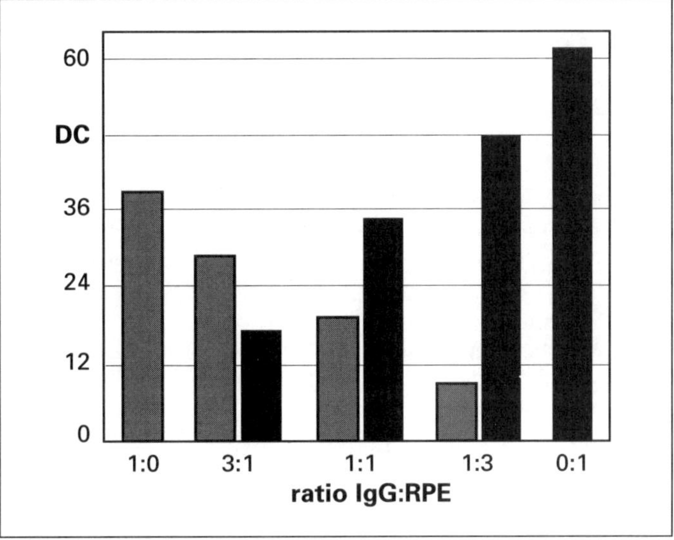

Figure 6.8. Reduction of IgG capacity from competition. Each data pair illustrates individual capacities for a mouse IgG_1 (gray) and R-phycoerythrin (black) when the 2 were applied simultaneously at the indicated mass ratio. Dynamic capacity values (DC) are indicated in mg/mL of gel.

samples applied to it, and to the methods and frequency of cleaning and sanitization. With advance delipidation and good column maintenance, large-scale columns exhibit the same longevity as small HPLC columns. A broad-spectrum sample treatment method is described in appendix II.

Limitations

Risk of antibody denaturation on excessively hydrophobic supports is the greatest concern with HIC. Strong HIC ligands provide more than sufficient energy to cause denaturation, and numerous studies have confirmed on-column protein-conformational changes.[59,81,83-87] This is a special concern with antibodies because their hydrophobic sites are either directly associated with maintaining structural integrity, or responsible for mediation of antibody effector functions.[88] Titer, pharmacokinetics, and stability are all at risk. Many on-column changes are reversed spontaneously on elution, but any indication of denaturation must be documented not to have modified the native performance characteristics of the antibody.

denaturation

Fortunately, denaturation problems are not an inevitable result of HIC per se, but an indication that a particular medium is too hydrophobic. Figures 6.9 and 6.10 illustrate this point and provide indicators of denaturation that can be used to qualify appropriate

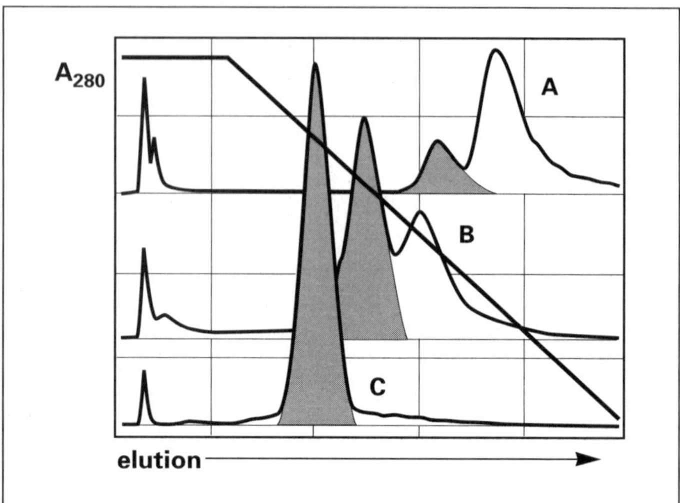

Figure 6.9. Antibody denaturation from excessive matrix hydrophobicity. Sample: 20µL purified mouse IgM. Antibody was eluted in a 10CV linear gradient from 1.5–0.0M ammonium sulfate at pH 7.0. The shaded areas indicate native antibody. Antibody collected from unshaded areas was denatured. A: a phenyl column. B: a moderately hydrophobic column. C: a weakly hydrophobic column. This particular antibody was unusually labile for even an IgM. Columns of mid-range hydrophobicity are applicable to most.

HIC media for a given monoclonal antibody. Phenyl ligands mark the upper boundary of acceptability for most IgG monoclonals, and are too hydrophobic for many. Among those for which phenyl media are appropriate, most require 25–50% ethylene glycol in the eluting buffer to obtain purity and recovery comparable to weaker ligands.

Ligands stronger than phenyl guarantee problems. Strong organic solvents in the mobile phase may improve mass recovery, but this practice is doubly dangerous. Denaturing influences are additive.[51,89] The combination of strong organic solvents with an excessively hydrophobic ligand maximizes the probability of permanent product denaturation.

variable selectivity among HIC media

It's important to appreciate that ligand name is not a reliable indicator of a support's hydrophobicity. The chemical identities of many products are obscured by trade names. Even among products for which the name accurately reflects the identity of the ligand, variations in ligand density, the length and hydrophobicity of the spacer arm, and the hydrophobicity of the base matrix all affect selectivity.[17-21,23,24,27,30,57,85,87,90-94] The only reliable way to classify column hydrophobicity is to screen media individually.

antibody solubility limitations

Most antibodies are partially or wholly insoluble at the salt concentrations required to support retention

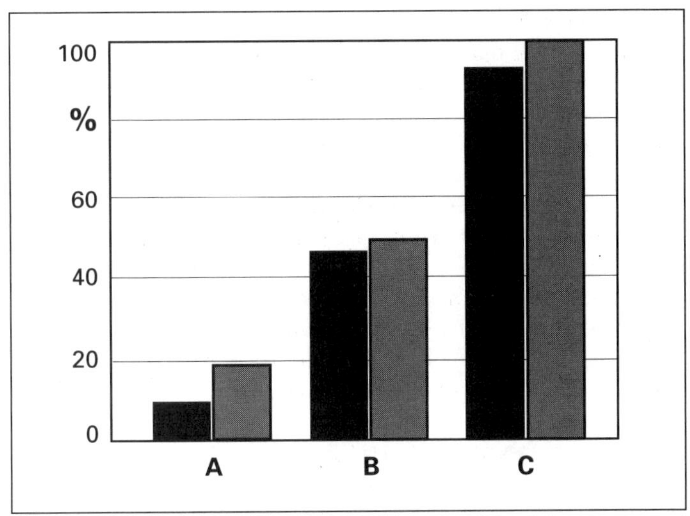

Figure 6.10. Antibody denaturation from excessive matrix hydrophobicity. This graph ilustrates mass and activity recovery from runs conducted on the same columns and under the same buffer conditions as Figure 6.9. Black bars indicate mass recovery. Gray bars indicate recovery of immunoreactivity per mg IgM. A: the phenyl column. B: the moderately hydrophobic column. C: the weakly hydrophobic column. All values expressed as % of native IgM in the pre-HIC sample.

on weakly hydrophobic media (Figure 6.11). This isn't a problem with injections of unequilibrated sample less than 2% of the column volume, but it's a serious limitation for preparative applications. Advance batch-equilibration is prohibited by formation of precipitates that can clog the column or otherwise affect the process.[81] Even if a sample appears soluble at the beginning of the load, aggregates formed over the duration of sample application may affect performance.

on-line dilution

Solubility limitations can be managed through loading sample by on-line dilution.[81] This technique exploits the time dependency of precipitation. By mixing the sample with a high-salt diluent in the chromatograph, immediately upstream from the column, average sample/salt contact time is reduced to the point where precolumn precipitation is no longer a limitation. Another advantage of this approach is that the sample/salt precolumn contact time is the same throughout the duration of sample application, regardless of flow rate or total sample volume.

On-line dilution is not a perfect solution, however. The first compromise is that it increases the process development burden. Loading buffer formulations vary with the antibody and the hydrophobicity of the chromatographic support. Fortunately, antibody solubility characteristics are farily uniform. This makes it possi-

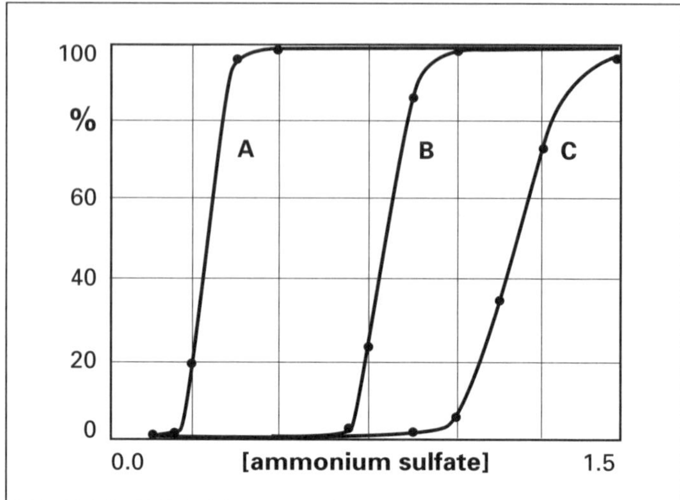

Figure 6.11. Hydrophobic interactions of Mouse IgG in ammonium sulfate. Curve A illustrates interactions with a phenyl column, B with a weaker column, and C illustrates antibody precipitation in free solution. For phenyl, retention is achieved at ~0.5M salt, well below the point at which precipitation commences. Sample can be equilibrated off-line in advance. Retention on the weaker matrix is achieved at ~1.0M salt. Antibody precipitation has begun to occur. On-line dilution is required for sample loading. See reference 81 for experimental details.

ble to apply generic conditions in early methd development, then refine them, rather than developing conditions for each antibody from scratch.[81]

Most antibodies will tolerate a maximum of ~2.0M ammonium sulfate or potassium phosphate in the binding diluent, and up to about 1.8M after dilution (Figure 6.12). Higher concentrations risk instantaneous precipitation of the antibody at the mixing interface. For moderately hydrophobic gels, the on-line dilution factor should be set so that the antibody is applied to the column in at least 1.2M salt. Higher salt concentrations support higher capacities and may be required to achieve antibody retention on weakly hydrophobic gels.

The second compromise is that flow cannot be stopped at any time during sample loading. Otherwise antibody will precipitate in the lines, creating the same problems as off-line batch equilibration. If creeping backpressure during a run warns that sample loading will have to be terminated prematurely, wash the protein out of the lines first.

other effects of concentrated salts

Directly or indirectly, high salt concentrations mediate other compromises, one of which is the elevated viscosity they impart to HIC binding buffers. This places greater constraints on an instrument's mixer efficiency.[28] It has also been shown to depress dynamic capacity by reducing protein diffusivity.[34]

Figure 6.12. HIC dynamic capacity as a function of salt concentration. This graph illustrates results for a purified mouse IgG_1 on a moderately hydrophobic support. The solid curve indicates dynamic capacity in mg/mL. The dashed curve indicates free solution turbidity in AU_{600nm}. The decrease in capacity coincident with the increase in turbidity shows that the column is unable to adsorb stable aggregates. They pass through the void volume. [82]

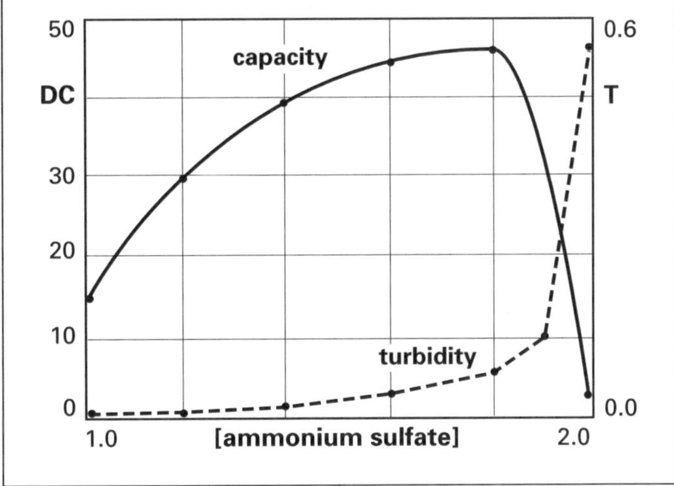

instrument wear

Increased wear on instrumentation is another factor. Even for "inert" systems, dried residual salts can cause irreparable scarring of pump and valve seals. For stainless steel systems, corrosion is an issue as well. This places greater constraints on instrument maintenance.

interference with analysis

High salt concentrations can affect analysis of column fractions, especially by electrophoresis. Salt causes gel dehydration at the point of sample application, and convective heating of the gel matrix. Both cause banding artifacts that confound interpretation. With IEF systems in particular, the heat can melt through the plastic gel backing and insulation on the cooling plate. This has led to catastrophic retirement of a surprising number of systems. Avoid this expensive and potentially hazardous lesson by equilibrating samples to more moderate salt concentrations. Multisample microdialyzers perform this task efficiently.

High salt concentration is generally not a concern for immunoassays because samples tend to be diluted to a level where salt effects are negligible. Neither is it a problem for analysis by protein A or protein G affinity chromatography. IgG binding to protein A is enhanced in high salt while protein G seems to be unaffected one way or the other (chapters 9 and 10).

Ammonium ions compete with amino groups in chemical modification reactions like those used for immobilization of antibodies on chromatographic supports or synthesis of immunoconjugates. They also interfere with total nitrogen assays.

disposal issues

Disposal of concentrated salt solutions can be a burden at large manufacturing scales. Sulfates and phosphates are often a particular issue with municipal authorities. Organic salts like monosodium glutamate offer some relief of this specific problem, but their higher procurement costs cancel out any real benefit.

metal contamination

Metal contamination of raw material binding salts is more problematical with HIC than other methods because the high salt concentrations translate into higher metal exposure for the proteins. Divalent metal cations bind to proteins, altering their net charge and reducing their stability.[2,3,95] This is a concern for IgGs because of the metal affinity of a highly conserved histidyl cluster at the juncture of the $C\gamma2$ and $C\gamma3$ domains.[63,64,96] Even part-per-million (ppm) metal contamination can be significant. For example, 1ppm nickel in a 1mg/mL solution of IgG corresponds to a molar excess of >50 fold. This reveals metal contamination as an important process variable.

The best place to control metal contamination is at the source. Obtain salts with the lowest possible metal contamination and log certificates of analysis to document consistency. Addition of EDTA is a worthwhile precaution but the ability of antibodies to bind immobilized nickel on IMAC columns illustrates that chelated metal in solution can bind just as effectively.[8] Addition of imidazole, histidine, or histamine will outcompete protein histidyl residues for available metal ions.[8,97-99] The liability with these agents is that they preclude IMAC for the subsequent processing step. EDTA also precludes HAC.[100]

temperature effects

HIC binding efficiency diminishes with temperature for the majority of proteins, including monoclonal antibodies.[60,61,85-87,90,91,101-104] Not only do temperature changes alter retention characteristics, they alter them to different degrees for different proteins. This warns that temperature must be carefully controlled to maintain reproducibility. Processes must

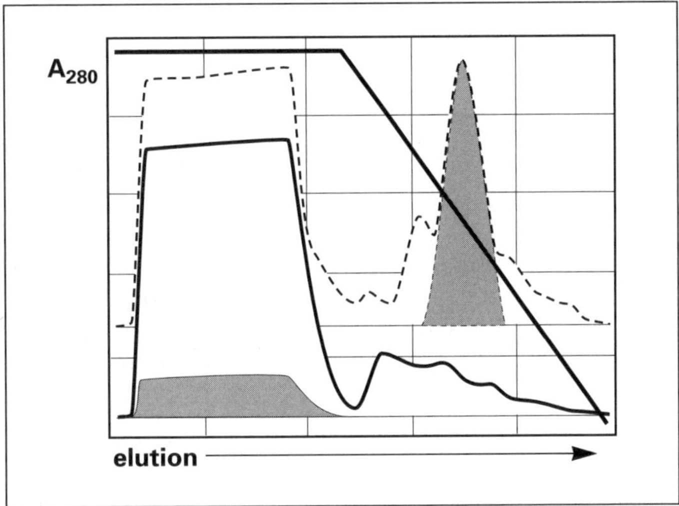

Figure 6.13. Process failure from uncontrolled temperature variation. The solid profile illustrates the result when sample was brought from 4°C and applied directly to a HIC column in a process designed for 21–23°. Dashed lines indicate the reference profile. Shaded areas indicate product distribution.

be developed at their eventual running temperature. Manufacturing SOPs should specify that all process materials be brought to temperature before initiating a process (Figure 6.13).[90]

Besides altering selectivity, reducing temperature from 22° to 4°C increases mobile phase viscosity by a factor of ~1.5.[105] This reduces protein diffusivity and increases pressure drop by a factor of ~2. Even if the system will tolerate the higher operating pressure, be prepared to reduce flow rate to compensate for lost separation performance and dynamic binding capacity.

A more insidious temperature issue is that some monoclonals, including up to 20% of IgMs, are cryoglobulins; progressively insoluble below 37°C.[106-110] HIC processes are seldom conducted in the cold, but it's important to be vigilant for these antibodies all the same. Unless samples are fully equilibrated to temperature before processing, there is a high risk of antibody remaining on the walls of the container instead of going onto the column. Cryoprecipitation is discussed more thoroughly in chapters 2–4.

buffer incompatibilities

A minority of strongly basic antibodies, or antibodies with strongly basic charge sites, can form stable crosslinks with polyvalent anionic salts such as phosphate and sulfate. These mainly include IgG_3s and some IgMs. This usually isn't a problem at the high salt

Table 6.3. Buffers and conditions for evaluation of column hydrophobicity. These conditions are appropriate for media less hydrophobic than phenyl. Representative ligand names include alkyl, ether, eth, iso, isopropyl, and propyl. Keep in mind that these names are not accurately descriptive of matrix hydrophobicity.

Buffers
A: 0.05M sodium phosphate, 1.5M ammonium sulfate, pH 7.0 **B**: A, minus ammonium sulfate
Conditions
Equilibrate column: with 10CV buffer A **Inject:** ≤2%CV unequilibrated sample **Wash:** 2CV buffer A **Elute:** in a 15CV linear gradient to buffer B. **Strip:** with a 5CV buffer B.

concentrations used for loading HIC columns but the antibody tends to form a fine precipitate upon elution. Removal of the precipitate is not a solution since the equilibrium restores itself spontaneously. In some cases the interaction can be suspended by titration with sodium or potassium fluoride.

Method development

Method development for HIC follows the same organizational pathway as IEC (chapter 4). Only points of difference are discussed here. The first priority is to qualify media that will not denature the product. If you have a control sample of purified antibody, media can be screened visually according to the shape of the eluted peak in a linear gradient. Screening conditions are suggested in Table 6.3. Use the peak configuration on the weakest medium as a control. Creation of additional populations on stronger hydrophobic media is cause for strong scrutiny, if not immediate rejection. Be aware that heterogeneity may be apparent even on the weakest medium due to pre-HIC degradation or aggregation of the antibody.[111,112]

media selection

Select the most hydrophobic medium that does not evidence denaturation.[81] The stronger the column, the lower the salt concentration at which your antibody will bind. For phenyl columns, the required salt concentration is often beneath the level at which antibody precipitation begins to occur. Off-line batch equilibration and direct sample loading are viable options. Another advantage: lower binding salt translates into lower salt in your eluted product fractions.

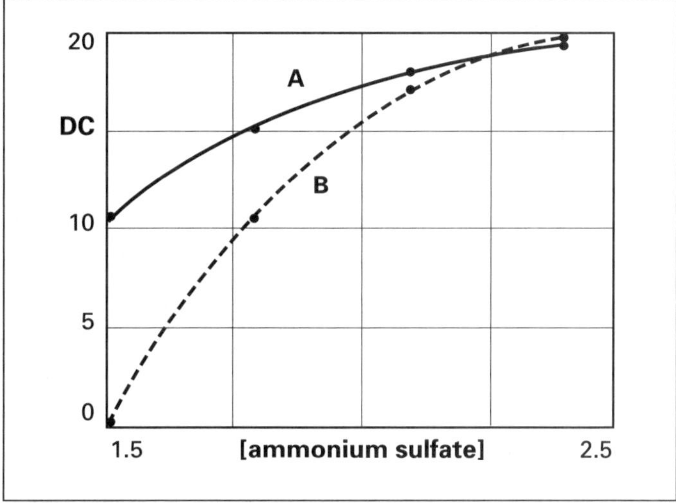

Figure 6.14. Dynamic capacity as a function of salt concentration on columns with different hydrophobicities. Sample: bovine serum albumin. Profile A is from a phenyl column. Profile B is from a weaker, moderately hydrophobic column. Both columns employ the same base matrix. Dynamic capacity is expressed in mg/mL of gel.

Perhaps the strongest rationale for using the most hydrophobic candidate is that reproducibility is better on stronger ligands.[82] Figure 6.14 compares dynamic capacity for bovine serum albumin on moderately and strongly hydrophobic columns as a function of salt concentration. Dynamic capacity changed by an average of 10.2% and 2.6%, respectively, per 0.1*M* ammonium sulfate over the range from 1.5*M* to 2.4*M*. Given that most external process variations simulate either an increase or reduction of binding salt, these results suggest that the more hydrophobic column should be ~4-fold less responsive to external variation.[90]

high-flow media

There are advantages to selecting the media that support the highest flow rates. Higher flow rates translate into shorter precolumn sample/salt exposure times. This miminizes the potential for precolumn precipitation no matter how you apply your sample. While speed is valuable, it doesn't come without a price. All supports lose capacity with increasing flow rate (Figure 6.7). The recent generation of fast chromatography media limits the proportional loss, but some achieve this benefit by using pore architectures that reduce inherent binding capacity to begin with.[113]

selectivity screening

After you've selected your medium, you can screen conditions for the most favorable selectivity. Evaluate potassium phosphate at pH 8.5, and potassium phosphate and/or ammonium sulfate at pH 5.5 and pH 7.0. Use the conditions listed in Table 6.3, except loading 2–5mg of antibody onto a 1mL column by on-line dilution: 80% buffer A, 20% sample.

evaluating selectivity

As with ion exchange it's important to evaluate contaminants in the context of their behavior on other candidate methods.[109] If one set of conditions yields antibody 30% contaminated with a protein that's easily removed by IEC, and another set of conditions yields antibody 5% contaminated with a protein that coelutes on IEC, then the former conditions are the better option. Evaluating the results according to percent purification is meaningless, except so far as the identity of the contaminants is the same.[34,114]

process order

HIC is well suited as a first purification step insofar as most contaminants are unretained under antibody-binding conditions. This provides maximum capacity and maximum concentrating ability for the product. It's also advantageous from the perspective of dissociating and removing complexed DNA early in the purification process. Vulnerability to lipid fouling is a weakness of HIC-first strategies. On-line dilution of already dilute products is also unattractive.

scale-up issues

Temperature is a more important scale-up parameter with HIC than with most other separation methods. Even if minor variations between the development and manufacturing areas don't result in a total scale-up failure, they will certainly amplify process variation. Synchronizing environmental conditions is worth the investment, whatever it takes.

For processes conducted on weakly to moderately hydrophobic media, the key scale-up concern is antibody solubility under sample application conditions. This is seldom a problem when sample is loaded by on-line dilution but it can be disastrous if the sample is equilibrated off-line in advance. Sample equilibration is less of an issue with stronger hydrophobic ligands but still bears careful evaluation.

An occasional problem with strongly hydrophobic supports is time-dependent on-column denaturation. Compare elution profiles from a run conducted within the intended duration, with a run in which the sample is left on the column for the worst-case duration. If the profiles differ in any way suggestive of denaturation, the best course is to convert to a less strongly hydrophobic column. Otherwise, scale-up issues with HIC are largely the same as with IEC.

In perspective

The speed, capacity, and resolving capabilities of HIC—especially concerning removal of DNA, viruses, endotoxin, and nonspecific antibodies—make it an indispensable processing tool. On the other hand, method development is burdened by the need to screen chromatography media for product denaturation, and by development of sample application conditions to compensate for product solubility limitations. Resolution of the conflict comes with accepting the technique for what it is, on its own terms. With practice, method development and manufacturing with HIC become as routine as with any other method.

Most IgG monoclonals for in vitro applications can be purified effectively with 2-step combinations of HIC with IEC, HAC or IMAC. HIC following cation exchange is attractive because it doesn't require intermediate sample equilibration. HIC following IMAC is even better because it doesn't require dilution of the raw sample to load the IMAC column. HIC/SEC combinations are effective for IgMs. All of these processes provide solid foundations for in vivo purifications.

Recommended reading

For detailed discussion of practical issues in preparative HIC, refer to the recent 4-part series in BioPharm by Gagnon et al [34,81,82,90]. For more general treatment, consult older reviews by Chicz and Regnier, by Kennedy, and by Eriksson[28-30]. References 2–5, 9 and 10 provide important refinements to earlier hypotheses concerning HIC retention mechanisms. Refer to the articles by Fausnaugh and Chicz for discussion of how distribution of protein surface features affect selectivity.[36,58]

References

1. A. Tiselius, 1948, Arkiv. för Kemi. Mineralogi. Geologi., **26B** 1
2. T. Arakawa and S. Timasheff, 1982, *Biochemistry*, **21** 6545
3. T. Arakawa and S. Timasheff, 1984, *Biochemistry*, **23** 5912
4. B. Roetger et al, 1989, *Biotech. Progr.*, **5** 79
5. T. Arakawa, 1986, *Arch. Biochem. Biophys.*, **248** 101

6. A. Galkin et al, 1984, *Anal. Biochem.*, **142** 252
7. T. Fujita et al, 1980, *J. Biochem.*, **87** 89
8. J. Porath and B. Olin, 1983, *Biochemistry*, **22** 1621
9. C. Helm et al, 1989, *Science*, **246** 919
10. J. Israelachvili and R. Pashley, 1982, *Nature*, **300** 341
11. Z. Er-El et al, 1972, *Biochem. Biophys. Res. Commun.*, **49** 383
12. R. Yon, 1972, *Biochem. J.*, **126** 765
13. B. Hoffstee, 1973, *Anal. Biochem.*, **53** 430
14. S. Shaltiel and Z. Er-El, 1973, *Proc. Nat. Acad. Sci.*, **70** 778
15. S. Hjertén, 1973, *J. Chromatogr.*, **87** 325
16. J. Porath et al, 1973, *Nature*, **245** 465
17. D. Gooding et al, 1984, *J. Chromatogr.*, **296** 107
18. H. Jennisen and L. Heilmeyer, 1975, *Biochemistry*, **14** 754
19. R. Srinivasan and E. Ruckenstein, 1980, *Sep. Purif. Met.*, **9**(2) 267
20. J.-P. Chang et al, 1985, *J. Chromatogr.*, **319** 396
21. D. Gooding et al, 1986, *J. Chromatogr.*, **359** 331
22. D. Nau, 1989, *BioChromatography*, **4** 62
23. J. Fausnaugh et al, 1984, *J. Chromatogr.*, **317** 141
24. Y. Kato et al, 1984, *J. Chromatogr.*, **298** 407
25. W. Melander and C. Horvath, 1877, *Arch. Biochem. Biophys.*, **183** 200
26. W. Melander et al, 1984, *J. Chromatogr.*, **317** 67
27. P. Strop, 1987, *J. Chromatogr.*, **294** 213
28. R. Chicz and F. Regnier, 1990, *Met. Enzymol.*, **182** 392
29. R. Kennedy, 1990, *Met. Enzymol.*, **182** 339
30. K.-O. Eriksson, 1989, in Protein Purification, Principles, High Resolution Methods, and Applications, (J.-C. Janson and L Rydén, eds.), p. 207, VCH, New York
31. —1990, Hydrophobic Interaction Chromatography, Principles and Method, Pharmacia Biotech AB, Uppsala, 32. V. Parsegan, 1995, *Nature*, **378** 335
33. F. Hofmeister, 1888, *Arch. Exp. Pathol. Pharmakol.*, **24** 247
34. P. Gagnon and E. Grund, 1996, *BioPharm*, **9**(5) 54
35. M. Schmuck et al, 1986, *J. Chromatogr.*, **371** 55
36. J. Fausnaugh and F. Regnier, 1986, *J. Chromatogr.*, **359** 131
37. S. Hjertén et al, 1986, *J. Chromatogr.*, **359** 99
38. H. Schultze and J. Heremans, 1966, Molecular Biology of Human Proteins, Vol. I, Elsevier, New York
39. L. Narhi et al, 1989, *Anal. Biochem.*, **182** 266
40. T. Arakawa and S. Timasheff, 1984, *J. Biol. Chem.*, **259** 4979
41. P. von Hippel and K.-Y. Wong, 1965, *J. Biol. Chem.*, **240** 3909
42. E. Kabat and M. Mayer, 1966, Experimental Immunochemistry, 2nd Ed., Charles C. Thomas Publisher, Springfield
43. J. Porath and N. Ui, 1964, *Biochim. Biophys. Acta*, **90** 324
44. U.-B. Hansson and E. Nilsson, 1973, *J. Immunol. Met*, **2** 221
45. A. Helenius et al, 1978, *Met. Enzymol.*, **56** 734
46. D. Hereld et al, 1986, *J. Biol. Chem.*, **261** 13813

47. J. Lee and S. Timasheff, 1974, *Biochemistry*, **13** 257
48. H. Inoue and S. Timasheff, 1968, *J. Am. Chem. Soc.*, **90** 1890
49. D. Robinson and W. Jencks, 1965, *J. Am. Chem. Soc.*, **87** 2462
50. F. Regnier, 1983, *Met. Enzymol.*, **91** 137
51. C. Tanford, 1968, *Adv. Protein Chem.*, **23** 121
52. S. Timasheff and G. Fasman, 1969, Structure and Stability of Biological Macromolecules, Marcel Dekker, New York
53. W. Jencks, 1969, in Catalysis in Chemistry and Enzymology, McGraw-Hill, New York
54. S. Timaseff and H. Inoue, 1968, *Biochemistry*, **7** 2501
55. S. Budavari (ed.), 1989, The Merck Index, 11th edition, Merck and Co., Rahway
56. M. Schmuck et al, 1984, *J. Liq. Chromatogr.*, **7** 2863
57. T. Miller and B. Karger, 1985, *J. Chromatogr.*, **326** 45
58. R. Chicz and F. Regnier, 1990, *J. Chromatogr.*, **500** 503
59. V. Zizkovsky et al, 1986, *Oncodev., Biol. Med.*, **3** 323
60. P. Strop et al, 1978, *J. Chromatogr.*, **156** 239
61. P. Strop and D. Chechova, 1981, *J. Chromatogr.*, **207** 55
62. S. Hjertén, 1973, *J. Chromatogr.*, **87** 325
63. J. Deisenhofer et al 1978, *Hoppe-Seylar's Z. Physiol. Chem.*, **359** 975
64. J. Deisenhofer, 1981, *Biochemistry*, **20** 2361
65. F. Saul et al, 1978, *J. Biol. Chem.*, **253** 585
66. C. Cantor and P. Schimmel, 1980, Biophysical Chemistry, Vol. I., Freeman, New York
67. R. Creighton, 1983, Proteins; Structure and Molecular Principles, Freeman, New York
68. D. Schmidt and F. Westheimer, 1971, *Biochemistry*, **10** 1249
69. C. Tanford, 1968, *Adv. Protein Chem.*, **23** 1
70. G. Sofer and L.-E. Nyström, 1991, Process Chromatography, A Guide to Validation, Academic Press, New York
71. M. Belew et al, 1994, *J. Chromatogr.*, **679** 67
72. D. Nau, *BioChromatography*, **5** 62
73. P. Gagnon, Special Weapons and Tactics for removal of Product-Bound DNA, slide presentation, BioEast '96, Washington, D.C.
74. R. Steiner, 1953, *Arch. Biochem. Biophys.*, **46** 291
75. R. Steiner, 1953, *Arch. Biochem. Biophys.*, **47** 56
76. D. Nau, 1982, in Nucleic Acid and Monoclonal Antibody Probes, (B. Swaminathan and G. Prakash, eds.) p.383, Marcel Dekker, New York
77. J. Grun et al, 1992, *BioPharm*, **5**(9) 22
78. M. Einarsson et al, 1981, *Virol. Met.*, **3** 213
79. P. Manucci and M. Columbo, 1988, *Lancet*, **2** 782
80. M. Weary and F. Pearson, 1988, *BioPharm*, **1**(4) 22
81. P. Gagnon et al, 1995, *BioPharm*, **8**(3) 21
82. P. Gagnon and E. Grund, 1996, *BioPharm*, **9**(3) 34
83. X. Geng et al, 1995, Thermodynamics of Protein Folding

on the Stationary Phase of High Performance Hydrophobic Interaction Columns, poster, 15th International Symposium on HPLC of Proteins, Peptides, and Polynucleotides, Boston
84. B. Hoffstee, 1975, *Biochem. Biophys, Res. Commun.*, **63**(3) 618
85. J. Rosengren et al, 1975, *Biochem. Biophys. Acta*, **412** 51
86. M. Kunitani et al, 1988, *J. Chromatogr.*, **443** 205
87. H.-L. Wu et al, 1986, *J. Chromatogr.*, **371** 3
88. D. Burton, 1985, *Mol. Immunol.*, **22** 161
89. S. Gerlsma and E. Stuur, 1974, *Int. J. Peptide Protein Res.*, **6** 65
90. P. Gagnon et al, 1995, *BioPharm*, **8**(4) 36
91. S. Miller et al, 1987, *J. Mol. Biol.*, **196** 641
92. J. Fausnaugh et al, 1984, *Anal. Biochem.*, **137** 469
93. M. Schmuck et al, 1986, *J. Chromatogr.*, **371** 55
94. P. Karsnas and T. Lindblöm, 1992, *J. Chromatogr.*, **599** 131
95. F. Gurd and P. Wilcox, 1956, *Adv. Protein Chem.*, **11** 312
96. J. Hale and D. Beidler, 1994, *Anal. Biochem.*, **222** 29
97. D. Rijken et al, 1979, *Biochim. Biophys. Acta*, **580** 140
98. I. Ohkubo et al, 1980, *Biochim. Biophys. Acta*, **616** 89
99. H. Kikuchi and M. Watanabe, 1981, *Anal. Biochem.*, **115** 109
100. T. Kawasaki, 1991, *J. Chromatogr.*, **544** 147
101. S. Goheen and S. Englehorn, 1984, *J. Chromatogr.*, **317** 55
102. A. Beggrund et al, 1994, *Process Biochem.*, **29** 455
103. T. Miller et al, 1984, *J. Chromatogr.*, **316** 519
104. R. Ingraham et al, 1985, *J. Chromatogr.*, **327** 77
105. L. Hagel, 1989, in Protein Purification; Principles, High Resolution Methods, and Appplications, (J.-C. Janson and L. Rydén, eds.) p. 63, VCH, New York
106. C. Middaugh and G. Litman, 1977, *J. Biol. Chem.*, **252** 8002
107. G. Litman et al, 1981, *Immunol. Commun.*, **10** 707
108. C. Middaugh et al, 1980, *J. Biol. Chem.*, **255** 6532
109. K. Weber and L. Clem, 1981, *J. Immunol.*, **127** 300
110. C. Middaugh et al, 1978, *Proc. Nat. Acad. Sci. USA*, **75** 3440
111. R. Harris et al, 1993, Structure and Heterogeneity of a Recombinant Antibody, poster, 13th International Symposium on HPLC of Proteins, Peptides, and Polynucleotides, San Francisco
112. R. Harris, 1996, Chromatographic Techniques for the Characterization of Human Monoclonal Antibodies, slide presentation, Waterside Monoclonal Conference, Norfolk,
113. P. Gagnon, 1996, *Valid. Biosys.*, **1**(2) 1, <http://www.validated.com/library.html>
114. P. Gagnon et al, 1993, *LC-GC*, **11**(1) 26

Chapter 7

Immobilized Metal Affinity Chromatography

"The Gods love the obscure and hate the obvious."
—Brihadaranyaka Upanishad

The origins of immobilized metal affinity chromatography (IMAC) date to the early 1960s but it didn't emerge as practical tool for protein purification until 1975.[1,2] The name IMAC is a collective term that has been proposed to include all types of solid-phase metal ion mediated interactions with proteins and other solutes.[3] It has been underexploited for monoclonal purification, but it's broadly applicable to IgGs and offers powerful process capabilities.[4,5]

Mechanism

Selectivity in IMAC is influenced by a wide range of variables, including the chelating group on the column, the immobilized metal ion, pH, identity and concentration of salt, and additives to either enhance binding or selectively elute bound proteins.[1-5] Fortunately, instead of degenerating into a labyrinth, these variables group naturally into a manageable number of systematic screening options.

The interaction of primary interest for monoclonal purification is the affinity of histidyl residues for chelated nickel. Mammalian IgGs have a highly conserved histidyl cluster at the junctures of the $C\gamma2$ and $C\gamma3$ domains.[5-8] The main concentration resides on the $C\gamma3$ domain. Under alkaline conditions, nickel-loaded iminodiacetic acid (Ni-IDA) columns support highly efficient and selective IgG capture through these residues (Figure 7.1).[3,5]

While histidyl residues support the strongest metal interactions, several others contribute to the associa-

Figure 7.1. IgG binding to immobilized nickel.

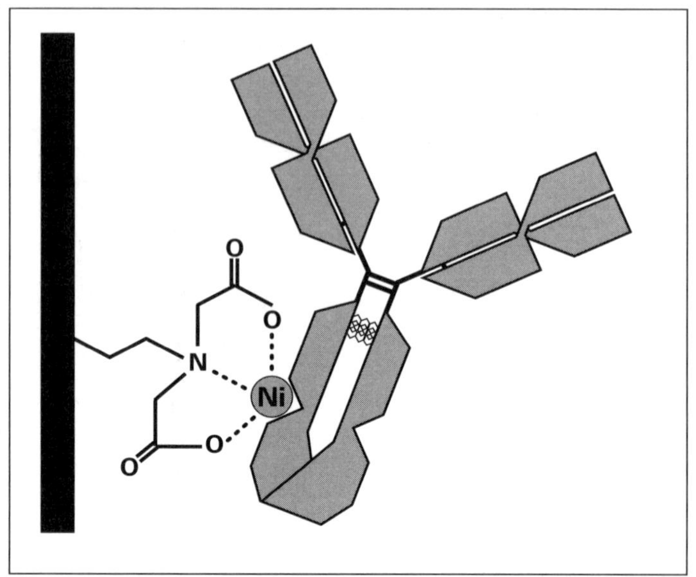

tion; in order: His > Trp > Tyr > Phe > Arg ~ Met ~ Gly.[3] One reference suggests that histidyl-binding specificity can be enhanced by including 0.5mM imidazole in the IgG loading buffer, and many media manufacturers have adopted this advice. This maintains the chelated nickel in its imidazolo form, specifically favoring histidyl interactions and theoretically improving reproducibility.[9] Other sources suggest that the benefits are negligible.[4] There is also the minor limitation, at least on some gels, that it prevents IgG binding (results not shown).

Ion exchange effects alter selectivity at low salt concentrations.[3,4] These effects are damped out at 0.1–0.5M sodium chloride. Selectivity is otherwise relatively salt-independent, even at concentrations up to 5.0M sodium chloride.[3,4]

elution

Elution can be achieved by reducing pH through the range of 8–4, mainly reflecting titration of the histidyl residues.[10] Competitive elution with imidazole or its analogues histidine and histamine yield similar selectivity (Figure 7.2).[3,4,11-13] A limitation with histidine and histamine is that the chelating capacity of their primary amino groups displaces metal:protein complexes from the IDA ligand. Elution can also be

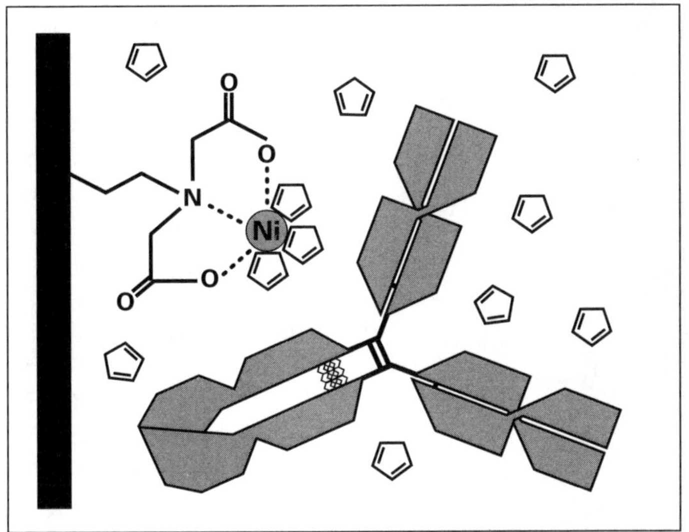

Figure 7.2. Competitive elution of IgG with imidazole. Note that the nickel remains bound to the column. IgG eluted by descending pH is similarly free of leached nickel.

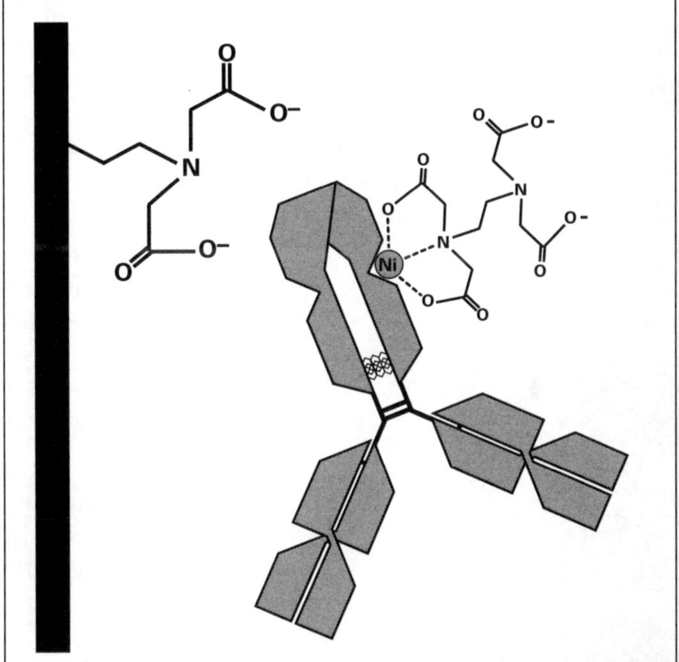

Figure 7.3. Elution of IgG-nickel complexes with EDTA. Note that the nickel remains bound to the antibody.

achieved with EDTA, which operates solely through displacement of metal from the column (Figure 7.3).

Like free amino acids, ammonium ions, amine-based buffers, and aminated zwitterionic detergents should be avoided in feedstocks and binding buffers;

citrate as well.[3,14,15] They reduce product binding capacity, may prevent binding altogether, and potentially contaminate the product with nickel. Cis-diol compounds are also weak chelators, interfering with binding and contributing to metal leakage.[3] Examples include ethylene glycol and sugars such as sorbitol and mannitol. Denaturing concentrations of urea generally prevent protein binding, but low concentrations are inert, as are nonionic detergents.[3,4]

Attributes

One of Ni–IDA's main strengths is the breadth of its applicability. IgG binding is sufficiently consistent that an IMAC kit is marketed as a generic method for removal of free enzyme from raw immunoconjugate mixtures.[16,17] The inability of a small fraction of Human polyclonal IgG to bind Ni-IDA columns predicts that some monoclonals may be unretained.[3] Circumstantial evidence suggests that the nonbinding population may be the G3m (s^-t^-) Caucasian allotype of IgG_3, in which Cγ3 histidine 435 is substituted by arginine.[18,19] This prevents protein A binding, which is also strongly histidine dependent (chapter 9).

purification performance

Data from experiments with polyclonal IgG in normal human serum suggest that IMAC should be capable of delivering total IgG purity of ~90%.[3] This has been confirmed with a humanized IgG monoclonal from cell culture supernatant. A mouse IgG from ascites at 0.1M sodium chloride was purified to only about 60%.[5] Increasing sodium chloride concentration to ≥0.5M may improve purity by 10–20%.

Ni-IDA's Fc specificity is a valuable asset with recombinant lines that overexpress light chain. A mouse/human chimeric IgG was purified successfully from a 50-fold light-chain excess.[5] Indiscriminant copurification of nonspecific IgG remains a limitation.

known protein contaminants

Other known contaminants include transferrin and traces of albumin.[5] Haptoglobin and C1q bind to Ni-IDA columns, but were apparently either absent from the media formulation or selectively fractionated by the elution conditions.[3,5]

DNA, endotoxin, and viral clearance

DNA, endotoxin, and viral clearance data have not been published. However, the ability of IMAC to tolerate loading at high salt concentrations supports dissociation of electrostatic contaminant:antibody complexes that plague low ionic strength methods. This improves DNA and endotoxin clearance, and enhances both product capture and separation performance.[20] In general, expect DNA, virus, and endotoxin clearance to rank with protein A (chapter 9).

mass and activity recovery

IgG mass recovery is reported to be >90%.[5] Activity recovery appears to be quantitative. Both findings are consistent with the mild operating conditions.

high capture efficiency

IMAC captures IgG efficiently from raw production media under near-physiological conditions.[4] Greater than 90% capture efficiency is obtained even with feedstream product concentrations as low as 1µg/mL.[5] This is on a par with the most avid bioaffinity ligands. The recent availability of IMAC media that support sample application without prior clarification maximizes the value of its product-capture ability.[21-24]

intermediate sample equilibration

Besides accommodating raw feedstreams, the relative independence of antibody binding from salt concentration offers a great deal of flexibility for process sequencing. IMAC can follow virtually any other technique without requirement for buffer exchange. The main exception is HIC conducted with ammonium sulfate. If you want to put HIC first, but want to avoid buffer exchange, use potassium phosphate as the binding salt. Mole for mole, its selectivity is similar to ammonium sulfate.[24]

IMAC can also be used to pre-equilibrate sample for downstream low ionic strength methods. High-salt loading improves IMAC binding specificity, but its concentration can be reduced after the major contaminants are washed through the column. Ion exchange effects will assert themselves below ~0.1M sodium chloride so be prepared to experiment. Even if you have to elute in salt as high as 0.1M, you'll still be within dilution range for most charge-based methods. This flexibility reveals IMAC as a valuable tool for streamlining overall process design. It offers nearly the same capability for linking steps as SEC, but does so without sacrificing speed or capacity.

Resistance to harsh sanitizing conditions

IMAC media tolerate treatment with concentrated acid, base, chaotropes, organic solvents, and detergents.[3,4] This allows rigorous cleaning and sanitization protocols without compromising column life. Nevertheless, the best approach is to not foul the column in the first place. IMAC is compatible with the advance foulant removal method in appendix II.

Limitations

Nickel is toxic in biological systems and will precipitate many antibodies. Studies have measured metal leakage from IMAC columns at ~6 parts per million.[25] These results were obtained with a copper-IDA column during elution of β-lactoglobulin with ammonium chloride at pH 7.2. Converting for nickel at the same concentration, and IgG at 5mg/mL, this corresponds to a 66 molar excess of metal. This raises concerns not only about toxicity and alteration of antibody performance, but also surface-charge alteration of the product.[26]

leached metal-ion control

Metal control is simple and can be integrated seamlessly into most processes. Elute with imidazole or by reducing pH. This suspends interaction of the antibody with the chelated metal while the nickel itself remains bound to the column. As a precaution, add EDTA with imidazole, histidine, or histamine to the column fractions. EDTA competitively blocks formation of coordination complexes between protein carboxyl clusters and divalent metal cations, while the imidazolium groups block histidyl/metal complexation. Including these agents in the binding buffer of the following process step ensures quantitative nickel removal.

Some references suggest the precaution of underloading IMAC columns with nickel to begin with.[3, 4,22] The rationale is that ions leached from the upper regions of the column will be scavenged by the lower regions. This is neither as simple nor as effective as it sounds. If you've eluted with anything other than imidazole or pH, at least some of your product will be metal-complexed. This means that the unloaded portion of the column must competitively dissociate the nickel from the antibody. Whereas the volume of unloaded gel needed to scavenge free nickel from solution is small, the volume required for quantitative competitive dissociation may be very large. Transfer kinetics must also be considered.

ion exhange effects

If you elute under low salt conditions, another limitation with the unloaded gel approach is that you risk secondary ion exchange interactions between the product and the unloaded IDA (cation exchange) groups. If you elute with imidazole or pH at >0.1M sodium chloride, scavenging columns can be effective. For best reproducibility set it up on a 2-column format, with a nickel-saturated purification column and a scavenging column 5% the volume of the purification column.

vulnerability to chelating agents in the sample

Chelating agents in the sample cause loss of IgG during column loading. Antibody binds to the stripped -chelated nickel and exits the column along with it (Figure 7.3). Imidazolium compounds leave nickel on the column but still block antibody binding. If losses are excessive, then use another technique for the capture step. Hydroxyapatite chromatography (HAC) may support antibody capture without significant sample dilution, but may also exhibit relatively low binding capacity due to competition between IgG and contaminants. Both cation exchange and HIC support efficient and more selective product capture, but often require significant sample dilution for loading.

desorption kinetics

Desorption kinetics may be slow in some instances, insofar as a second identical elution step will recover product that was not eluted on the first pass.[3] This has not been reported for antibodies, but you should evaluate the possibility during method development by plotting recoveries versus elution flow rate. This is especially important if you intend to run step gradients.

variability among chromatography media

Ligand density is an important determinant of selectivity and capacity, which vary substantially among commercial IMAC products.[3,22,27] Promotional literature indicates that tentacle-type media may exhibit twice the capacity of agarose gels and require 3 times as much imidazole for elution.[27] Expect dynamic capacities of 10-20mg IgG per mL of gel and imidazole elution molarities of 15 to 45mM. Antibodies elute at ~pH 6.0 regardless of column variations.

Table 7.1. Buffers and screening conditions. The sample equilibration step is optional, but the elevated pH improves capacity, and the high salt helps to dissociate ionic complexes between antibodies and anionic foulants like DNA and endotoxins. If IMAC is to be followed by a charge-based separation step, wash the column after sample loading with 0.05M potassium phosphate, 0.1M sodium chloride, pH 8.0. Elute as indicated, but substituting 0.1M sodium chloride for the 0.5M salt in buffer C.

Buffers

A: 0.25M potassium phosphate, 2.5M sodium chloride, pH 8.5
B: 0.05M Tris, 0.5M sodium chloride, 0.05M EDTA, pH 8.0
C: 0.1M sodium acetate, 0.50M sodium chloride, pH 4.5
D: C + 0.1M nickel chloride
E: 0.05M potassium phosphate, 0.5M sodium chloride, pH 8.0

Conditions

Equilibrate sample: add 1 part buffer A to 4 parts sample.
Strip column: with 2CV buffer B.
Wash: with 2CV buffer C.
Charge column: with buffer D until bed is uniformly colored throughout or until UV absorbance at 400nm reaches a plateau.
Wash: with 5CV buffer C.
Equilibrate column: with 5CV buffer E.
Load: up to 10mg IgG per mL of gel.
Wash: with 2CV buffer E.
Elute: in a 10CV linear gradient to buffer C.

inconvenience of multiple buffers

The extra steps of stripping and recharging nickel for every run can be tedious during method development. If you are eluting with imidazole, you may have no choice, since residual imidazole may block IgG binding in subsequent runs. Otherwise, experimental results document no change in separation performance over the course of 20 runs without recharging the column with metal.[4] When it comes to scale-up and manufacturing, the column should stripped and charged freshly for each lot of product nevertheless.

buffer incompatibilities

Some strongly basic antibodies form stable ionic complexes with polyvalent anions such as phosphate. This is encountered most often among IgG_3s. The consequent aggregation compromises both purification and recovery, regardless of the separation mechanism. Unfortunately, all the alternative buffers with pKs appropriate for IMAC are amine-based.

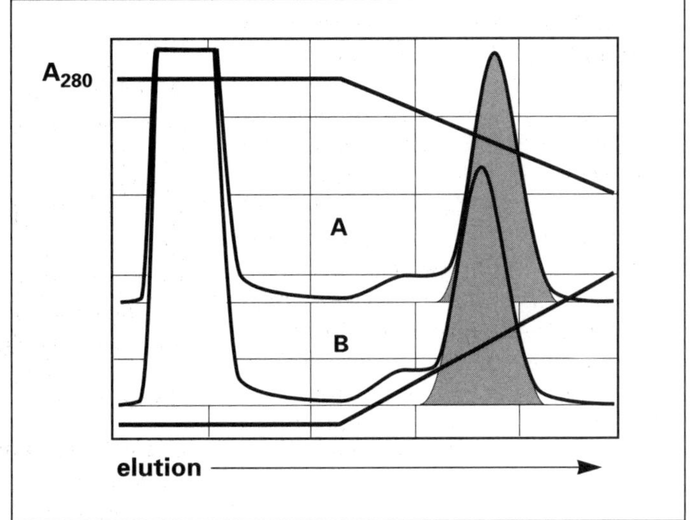

Figure 7.4. Elution profiles of mouse ascites with descending pH elution (A) and increasing imidazole elution (B). The selectivities are essentially indistinguishable. They can be brought into near-perfect confluence by adjusting the gradient intervals.

One option is to stay with phosphate but add sodium fluoride to dissociate the complexes. To determine the effective concentration, complex a sample of purified antibody with 0.05M phosphate at pH 8.0. The solution will be hazy and may contain visible precipitates. Titrate with sodium fluoride until the solution achieves clarity. IMAC has been conducted successfully in 0.1M Tris despite its amine-based structure, so the technique need not be abandoned if the fluoride approach fails.[3] However, use of an amine-based buffer should be accompanied by careful evaluation of metal leakage, binding capacity, and reproducibility.

Method development

Method development for IMAC follows the same general guidelines as IEC (chapter 4). Only points of difference will be discussed here. Table 7.1 lists conditions for initial screening (Figure 7.4). pH gradient elution is better suited at this stage, mainly because antibodies elute at about the same pH regardless of variations among columns. Imidazole elution requires an additional series of experiments to identify an appropriate working range. Imidazole also absorbs UV at 280nm, while pH elution gives a flat baseline. A final limitation is that residual imidazole may require nickel-stripping and recharging after each run.

process order

IMAC is usually best placed first in the process sequence, for 3 reasons: sample can be bound with lit-

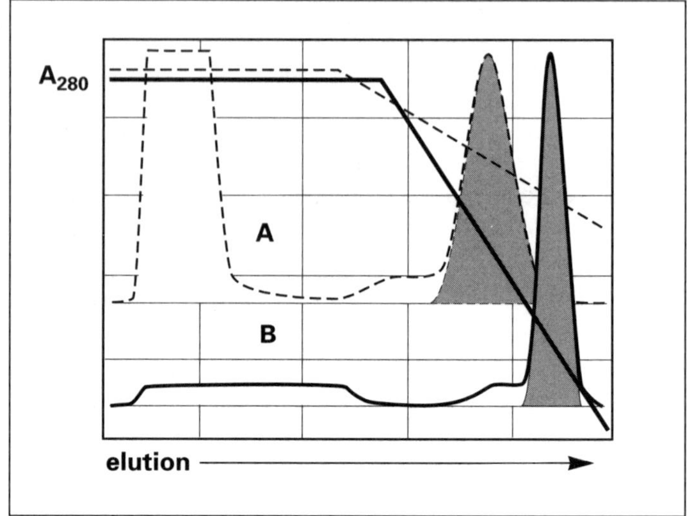

Figure 7.5. 2-step purification of mouse IgG from ascites by IMAC (A) and HIC (B). The IgG-containing fraction in each chromatogram is indicated by the shaded area. The leading shoulder on the HIC peak is polyclonal mouse IgG. See appendix I, protocol 5 for application details. Both methods are easily adapted for step elution.

tle or no equilibration, the majority of contaminants pass through the column, and downstream steps can remove any leached metal. The major exception is if the sample contains chelating agents, in which case it will be necessary to precede IMAC with another technique. Consider HAC, cation exchange, and HIC. If you use HIC, use a nonmmonium binding salt.

scale-up issues

The main scale-up points to watch with IMAC are the degree to which the column is metal saturated, the chelating agent content of the sample, and potential leached nickel interactions with the product. Otherwise, IMAC is vulnerable to the usual hazards for adsorption chromatography (see chapter 4).

In perspective

The authors of a 1983 article lamented that IMAC was slow to enter the mainstream of purification despite its "extraordinary merits".[3] They would be gratified to know that it's finally achieving recognition as the catalyst and nucleus for a generation of IgG purification procedures that are redefining the standards for process performance and economics.

IMAC vs bioaffinity

IMAC nullifies most of the rationale for using biological affinity methods. It binds IgG over a wide range of salt concentrations with little or no sample preparation. It binds it from more species and subclasses and with more uniform capacity than either protein A or protein G. It operates under far milder conditions. It requires less method development and provides better IgG specificity than protein A. It doesn't leach cytotoxic biological ligands into the product. It's less expensive, and it's much more durable.

purification of in vitro products

Single-step IMAC purifications may be adequate for antibodies produced by in vitro cell culture, but ascites-produced monoclonals require an additional step to remove nonspecific IgG and other contaminants. Two powerful combinations are IMAC followed by HIC and IMAC followed by cation exchange (Figure 7.5).[5] Neither require dilution of the raw sample. Both can be conducted without intermediate sample equilibration. The sequences can be reversed if there are chelating agents in the raw feed. Chromatographic materials are available that allow these combinations to be completed in 1–2 hours—at any process scale.

purification of in vivo products

Both combinations make excellent foundations for purification of in vivo products. IMAC's ability as an intermediate buffer re-equilibration tool makes it even more valuable. As its DNA, virus, and endotoxin clearance abilities become better appreciated, expect it to become the cornerstone of IgG purification for in vivo applications.

Recommended reading

The Chapter by Kågedal is a good point of entry into the IMAC literature.[4] The article by Hale and Beidler describes purification of IgG monoclonals.[5]

References

1. F. Helferich, 1961, *Nature*, **189** 1001
2. J. Porath et al, 1975, *Nature*, **258** 298
3. J. Porath and B. Olin, 1983, *Biochemistry*, **22** 162
4. L. Kågedal, 1989, in Protein Purification; Principles, High Resolution Methods, and Applications, (J.-C. Janson and L. Rydén, eds.), p. 227, VCH, New York
5. J. Hale and D. Beidler, 1994, *Anal. Biochem.*, **222** 29
6. *J. Diesenhoefer et al, 1978, Hoppe-Seylar's Z. Physiol. Chem.*, **359** 975
7. D. Burton, 1985, *Mol. Immunol.*, **22** 161
8. E. Kabat et al, 1987, Sequences of Proteins of Immunological Interest, US Dept. of Health and Human Services, Public Health Service, National Institute of Health
9. E. Sulkowski, 1985, *Trends Biotechnol.*, **3** 1
10. E. Sulkowski, 1987, in Protein Purification: Micro to Macro, UCLA Symposium on Molecular and Cellular Biology New Series, Vol. 68, (R. Burgess, ed.), p. 149, Alan R. Liss, New York

11. D. Rijken et al, 1979, *Biochim. Biophys. Acta*, **580** 140
12. I. Ohkubo et al, 1980, *Biochim. Biophys. Acta*, **616** 89
13. H. Kikuchi and M. Watanbe, 1981, *Anal. Biotechnol.*, **115** 109
14. E. Lance et al, 1983, *Anal. Biochem.*, **133** 492
15. D. Perrin and B. Dempsey, 1974, Buffers for pH and Metal Ion Control, Chapman and Hall, New York
16. FreeZyme, Pierce Chemical Company Rockford, IL USA
17. D. Boorsma and G. Kalbeek, 1975, *J. Histochem. Cytochem.*, **23**(2) 200
18. D. Haake, 1982, *J. Immunol.*, **129** 190
19. S. Ito et al, 1980, *Proc. Jpn. Acad.*, **56** Ser. B:226
20. P. Gagnon, 1996, Special Weapons and Tactics for Removal of Product-Bound DNA, slide presentation, BioEast '96, Washington, D.C.
21. TC Chelating CellThru BigBeads, Sterogene Bioseparations, Carlsbad, CA USA
22. Prosep-Chelating-I, Bioprocessing Ltd., Consett, UK
23. STREAMLINE-chelating, Pharmacia Biotech AB, Uppsala
24. P. Gagnon and E. Grund, 1996, *BioPharm*, **9**(5) 54
25. Prosep-Chelating-I, Product Data Sheet, Bioprocessing Ltd., Consett, UK
26. F. Gurd and P. Wilcox, 1956, *Adv. Protein Chem.*, **11** 312
27. W. Müller et al, 1994, Proven Tentacle Phases for Protein Separations, poster, PrepTech '94, Seacaucus, NJ USA, reprint: EM Separations Technology, Gibbstown, NJ USA

Chapter 8

Other Physicochemical Chromatography Methods

"If the left one don't getcha' then the right one will."
—Sugar-Ray Robinson (attributed)

Promising new antibody purification techniques like hydrophilic interaction chromatography (HILIC) are continuing to emerge and may soon enter the mainstream of monoclonal purification. A variety of other physicochemically-based techniques have been available for over a decade. Some are valuable complements to more widely used methods. Others are functional analogues. The remainder are technical curiosities.

Hydrophilic interaction chromatography

The term hydrophilic interaction chromatography was coined only recently: 1990, even though its origins date back over 30 years.[1-10] For large proteins, HILIC is the chromatographic analog of precipitation with polyethylene glycol (PEG). Proteins are adsorbed to a highly-hydrated uncharged adsorbent and eluted in a descending PEG gradient. The technique is just beginning to be systematically characterized, but it's already apparent that it offers unique and valuable process attributes.

The mechanism by which HILIC adsorption occurs is related to the mechanism for hydrophobic interaction chromatography (HIC). Like lyotropic salts, PEG is excluded from protein and chromatography media surfaces, resulting in their preferential hydration.[11-14] As PEG concentration increases, it becomes energetically more favorable for proteins to share their hydration shells with the chromatography support rather than remain individually soluble (Figure 8.1).[14] Up to this point, the mechanism is function-

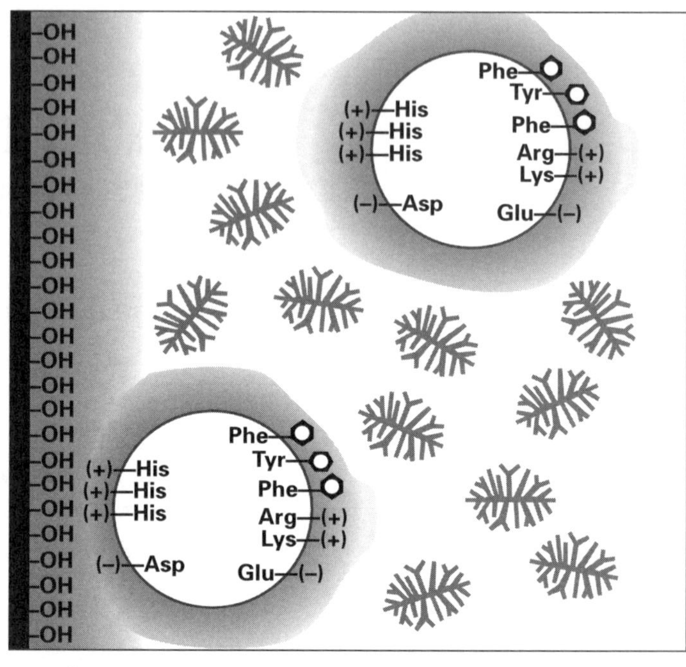

Figure 8.1. PEG-mediated protein adsorption by cohydration/cosolvent exclusion. The adsorbed protein is oriented with its most hydrated surface toward the gel. Hydrophobic surfaces are more weakly hydrated than polar surfaces. This partly explains why adsorption fails to occur on hydrophobic supports (Figure 8.2).

ally equivalent to the salt-mediated cohydration/cosolvent exclusion component of HIC (chapter 6).

HILIC vs HIC

HILIC's distinction is that selectivity is solely a function of protein and matrix hydration. Hydrophobic interactions do not make a significant contribution, even with columns that bear strong hydrophobic ligands. This is illustrated in Figure 8.2 by the behavior of an IgG in PEG on HIC supports with 2 different hydrophobicities. It was impossible to bind the antibody at PEG concentrations up to twice those required to precipitate IgG from solution. At 3 times the threshold concentration, a sufficiently stable precipitate was formed that it was trapped physically on the column. Note that the antibody "eluted" at the same PEG concentration despite the different column hydrophobicities, in both cases exiting the column as a milky suspension. This is reflected in the peaks' breadth and jaggedness. These points confirm that the antibody neither bound nor eluted in the usual sense. When PEG concentration descended to a point where the precipitates began to destabilize, the subaggregates simply washed through the column.

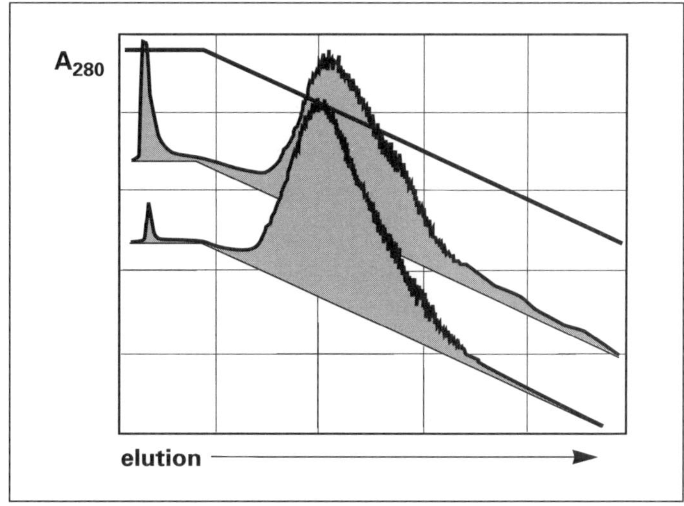

Figure 8.2. Atypical "elution" of mouse IgG on hydrophobic column in concentrated PEG. Sample: 20µg purified IgG in phosphate buffered saline was injected onto 1mL columns. Profile A was developed on a phenyl column; profile B on a less hydrophobic column. The slanted baselines are due to refractive index changes over the course of the PEG gradient: 20% to 0% in 20 column volumes.

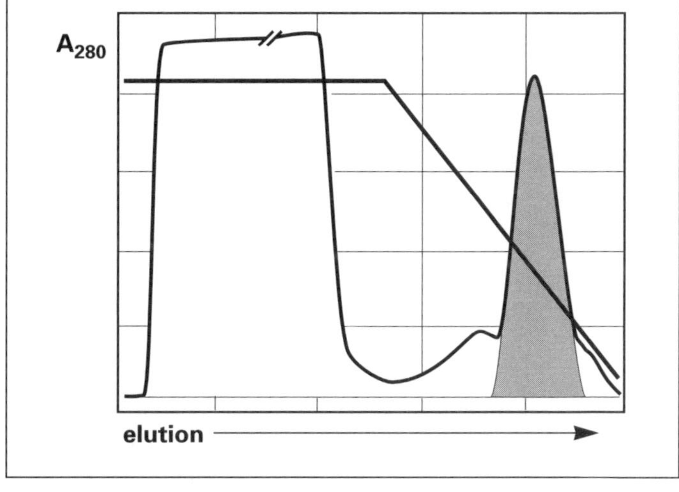

Figure 8.3. HILIC of a mouse IgM from ascites. Sample was loaded by on-line dilution at a ratio of 1 part ascites to 9 parts binding buffer. The binding buffer was 10% PEG-6000, 0.05M sodium phosphate, 0.10M sodium chloride, pH 7.0. The column was eluted in a 15CV gradient to 0% PEG.

attributes

Because of the root similarity of their mechanisms, selectivity on HILIC is similar to HIC on weakly hydrophobic supports. Expect purity ranging from 60–90%. Figure 8.3 illustrates a typical profile of an IgM fractionated from ascites by HILIC. Albumin and transferrin pass through the column while the antibody is strongly retained, eluting late in the gradient.

With current chromatography supports for HILIC, the technique is best suited to purification of IgMs, most of which adsorb at PEG concentrations less than 10%. IgG monoclonals mostly require at least 15%

PEG and some still fail to bind. This behavior parallels selectivity in PEG precipitation.[15] Antibodies elute at 30–50% of the PEG binding concentration (Figure 8.4).

mass and activity recovery

Current estimates of mass recovery range from 80–90%. Recovery of immunoreactivity per mass unit has been quantitative. This highlights an important advantage over HIC. Excessively hydrophobic supports are documented to denature antibodies.[16] Lack of hydrophobic interactions in combination with PEG's inherent protein-stabilizing properties makes HILIC ideally suited to purification of labile antibodies—especially IgMs.[12-14,17-19]

compatibility with downstream methods

Another important advantage over HIC is that you can proceed from HILIC to a low-salt charge-based second step without intermediate buffer exchange. The product pool from Figure 8.3 was applied directly to a cation exchanger after pH titration. Residual PEG does not compromise IEC binding; it augments it.[20] This reflects cooperativity between the cohydration and ion exchange mechanisms. Adsorption is similarly enhanced with hydroxyapatite (HAC), thiophilic adsorption, and protein A affinity chromatography.[21-23] Following HILIC with IEC is an effective means for removing residual polymer. It's unretained. Simply wash it through.

limitations

The primary restriction on HILIC is the current lack of chromatography media designed specifically for its exploitation. Such media are currently manufactured for peptides but not for proteins.[24] The media used for all the figures in this chapter is marketed as a weak HIC ligand: oligo-polyethylene glycol.[25] That immobilized PEG can be applied to both techniques reflects its dual character. At the high salt concentrations used for HIC, PEG is excluded from aqueous solutions. This reflects its hydrophobic aspect. Despite being a nonionic organic polymer, it is highly hydrated in low salt solutions. This is indicated by its high aqueous solubility: >30% w:v. This is a favorable indicator for the future of HILIC. For a ligand with the intermediate character of PEG to be able to support

Figure 8.4. PEG concentration at elution of IgG and IgM monoclonal antibodies. All IgGs were IgG_1.

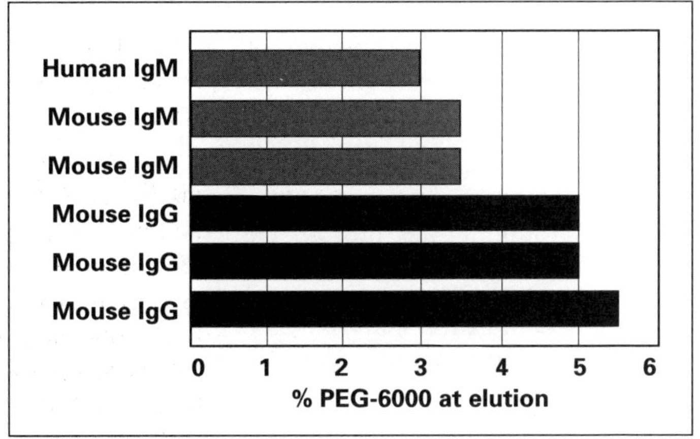

HILIC at all, means that development of supports optimized for the technique are likely to expand its scope of applicability.

high backpressure

HILIC is also limited by the viscosity of the PEG solutions required to promote adsorption. Dextran solutions have the same limitation. Gels with high mechanical strength are required to support the elevated backpressure, and high porosity is required to compensate for the loss of protein diffusivity. Hopefully the development of HILIC-optimized chromatography media will allow reduction of mobile phase polymer concentration, and possibly expand the scope of promoters to include less viscous—but still nondenaturing—alternatives.

antibody solubility limitations

Advance off-line sample equilibration is precluded by the insolubility of antibodies at the PEG concentrations required to achieve retention. This limitation is easily overcome by conducting sample equilibration and loading by on-line dilution (chapter 6).

removal of residual PEG

PEG-6000 behaves hydrodynamically like a globular protein of 50–100kD. This precludes its removal by buffer exchange chromatography, dialysis or diafiltration. It is barely separated from IgG by high resolution SEC, although easily removed from IgMs. Lower molecular weight PEGs can make size-based removal methods more effective but they must be applied at much higher concentrations to compensate for their lower molar effectivity in promoting adsorption.

Table 8.1. HILIC buffers and screening conditions. These conditions are appropriate for most IgMs. Use 15% PEG-6000 in Buffer A for screening IgGs. If you know the pI of the antibody, adjust buffer pH to match. Otherwise screen the 3 pH values at right to determine which supports the best antibody binding and selectivity. To generate sufficient sample for analysis, load 1–5mg antibody per mL of gel by on-line dilution at 1 part sample to 9 parts buffer A.

Buffers

A1: 0.05 sodium phosphate, 0.1M sodium chloride, 10% PEG-6000, pH 8.0

B1: A1 minus PEG

A2: 0.05 sodium phosphate, 0.1M sodium chloride, 10% PEG-6000, pH 7.0

B2: A2 minus PEG

A3: 0.05 sodium citrate, 0.1M sodium chloride, 10% PEG-6000, pH 6.0

B3: A3 minus PEG

Conditions:

Equilibrate column: 5 CV buffer A

Load: ≤2% CV unequilibrated sample

Wash: with 2–5CV buffer A

Elute: with a 10–15CV linear gradient to buffer B

Strip: with 5CV buffer B

interference with assays

High PEG concentrations interfere with electrophoretic separations by dehydrating the gel at the point of sample application. This isn't a problem with eluted antibody but it complicates evaluation of unretained fractions. For a qualitative estimate of contaminant content, precipitate the sample with ammonium sulfate and resuspend the precipitate with an appropriately low ionic strength buffer.[26] For a quantitative estimate of antibody content, substitute affinity chromatography for electrophoresis. PEG phase separation may cause problems for protein A columns equilibrated to high salt concentrations, but protein G and anti-light chain ligands are unaffected.[23]

gradient monitoring

PEG concentration in elution gradients cannot be monitored spectroscopically, by pH, or by conductivity. The magnitude of the gradient also precludes the use of refractive index detectors. However, it can be tracked easily with a pressure trace.

method development

Process development follows the same sequence as IEC (chapter 4). Table 8.1 describes screening conditions. The major scale-up limitation at present is the sparse selection of process chromatography media.

HILIC in perspective

HILIC is already being used in combination with IEC for commercial purification of antibodies for in vitro diagnostic applications. The ability to proceed from HILIC to IEC without intermediate buffer exchange is attractive from a process economy perspective, as well as for the high compound purification it offers. Media manufacturers will hopefully catch up quickly to the technique and help it fulfill its potential.

Euglobulin Adsorption Chomatography

Euglobulin adsorption chromatography of IgM is usually misclassified as a variant of size exclusion chromatography (SEC). The separation mechanism has nothing to do with size exclusion; it merely uses SEC media as an adsorbent. The technique takes advantage of the poor solubility of IgMs at low ionic strength. When applied to SEC media equilibrated under these conditions, soluble contaminants flow through the gel. The IgM adsorbs and subsequently desorbs at the advancing high salt buffer front.[27-29]

attributes

This method works with ~90% of monoclonal IgMs.[27] Purity from raw feedstreams is about 90%. The method's hallmark is that it supports quantitative removal of α_2-macroglobulin, which contaminates IgM preparations purified by many methods, particularly conventional SEC.[27-29] Preparations have also been reported to be free of protease activity.[29] Mass recovery ranges from 50–80%.[27] Recovery of immunoreactivity is apparently quantitative.

limitations

The SEC media should be equilibrated with a buffer of ~1000mS conductivity. Lower ionic strength risks antibody precipitation in the column.[27] Originally reported elution buffers employed very high conductivity: >1.5M sodium chloride. Most IgMs elute well beneath this level. If you want to miminize eluting salt concentration to facilitate a subsequent IEC or HAC step, try 0.05M HEPES, 1.0M urea and 1.0M glycine, 0.1M sodium chloride, pH 7.0. Urea disrupts the hydrogen bonds known to be involved in self-association of IgMs.[30] The risk of denaturation at 1.0M is nil.[31-33] Glycine solubilizes and stabilizes antibodies, especially IgMs.[34-36] Since urea is nonionic and

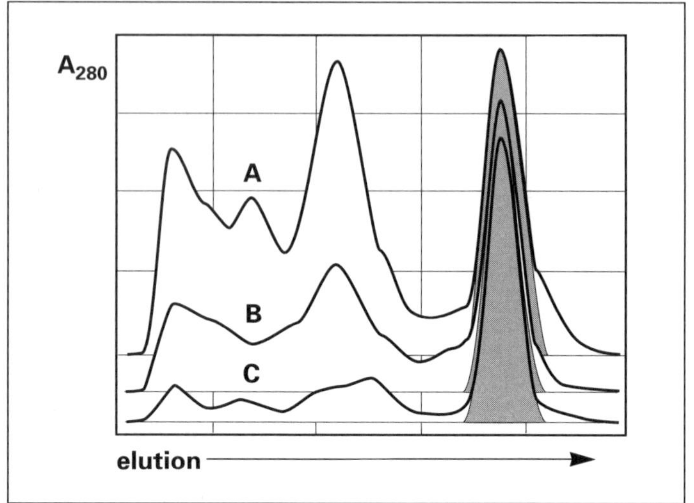

Figure 8.5. Euglobulin adsorption /desorption profiles of an IgM. Profile A: raw IgM ascites. Profile B: The same antibody after an anion exchange first-step. Profile C: The same antibody after a cation exchange first-step.

glycine is zwitterionic, neither interfere with charge-based separation methods.

The original applications of euglobulin adsorption employed the long narrow columns typical of SEC. However, bed heights >15cm are counterproductive. Results vary among SEC media but all seem to work.[27-29] Interestingly, gels with exclusion limits beneath IgM's mass are as effective as gels with higher exclusion limits.[27] Two-step combinations with charge-based separation methods yield preparations with purity suitable for investigational or in vitro applications.[28.29] Typical results are illustrated in Figure 8.5. A detailed protocol is provided in appendix I.

Thiophilic adsorption chromatography

Thiophilic adsorption chromatography was introduced in 1985.[37] The original "T-gel" was synthesized by coupling mercaptoethanol to divinylsulfone-activated Sepharose (Figure 8.6). The incorporation of the sulfone and thioether groups into an otherwise straight alkyl chain was believed to create a unique specificity for antibodies.[37-45] Subsequent studies have shown that replacing the sulfur groups with oxygen or nitrogen doesn't significantly alter selectivity, and in practice, the technique has proven to be a functional analogue of HIC.[22,41]

specificity

Early reports that thiophilic adsorption provided better specificity than HIC were confounded by the

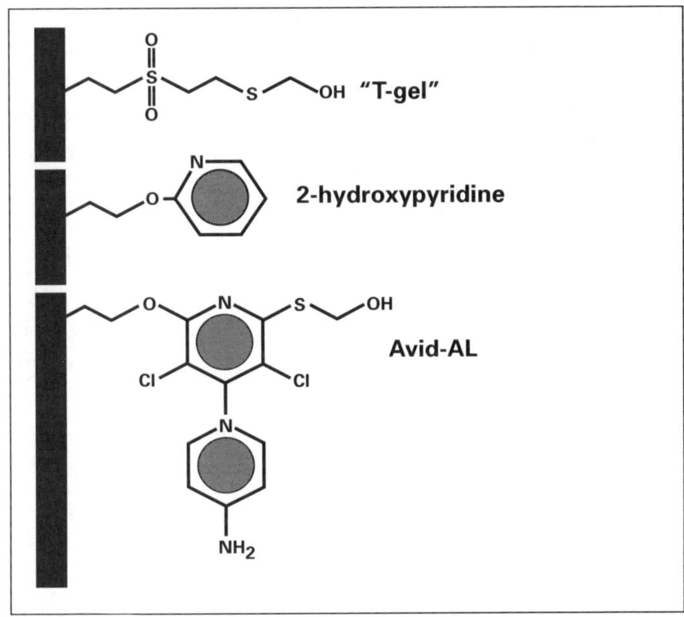

Figure 8.6. Ligands for "thiophilic" adsorption chromatography.

excessive hydrophobicity of the octyl media against which T-gel was compared.[37-43] T-gel itself is less hydrophobic than octyl media but still too strongly hydrophobic for commercial purification of most antibodies. This is indicated by low recoveries in comparison with HIC ligands of appropriately low hydrophobicity, and especially by denaturation of IgMs.[41]

thiophilic variants

Substitution of pyridyl rings for the straight-chain configuration of the original T-gel also supports efficient antibody binding in high salt.[22] Thiopyridyl ligands have been commercialized for purification of IgG and IgM.[46] Promotional literature claims product purity and recovery >90% at capacities of 10mg per mL of gel. Binding conditions are typical of strong hydrophobic ligands: 0.5–0.6*M* ammonium sulfate.

Elution is achieved by reducing the polarity of the mobile phase with 60% ethylene glycol. This is within the range where ethylene glycol is known to induce denaturation, highlighting the denaturative risks associated with the technique.[33,47] Ethylene glycol concentration can be reduced somewhat by inclusion of triethylamine or TRIS up to 1.0*M*.[48] These and other strong electron donors are thought to disrupt coplanar hydrophobic stacking of the ligand with aromatic residues on the protein.

Avid AL

Avid AL represents another variation, based on a thiodipyridyl compound.[48-53] The elevated hydrophobicity from the double-ring structure allows it to capture antibody under physiological conditions, but the risk of product denaturation is increased at the same time. This is a serious problem when elution is conducted at pH 3.0. Although sample loading can be conducted at physiological pH and conductivity, capacity is higher when samples are applied in ~0.5M ammonium sulfate. Denaturative losses are moderated by elution at neutral pH in the presence of polarity reductants and electron donors as discussed above.

thiophilic adsorption versus bioaffinity

Thiopyridyl and thiodipyridyl ligands provide several advantages over protein A and protein G. They bind more antibody classes from a wider variety of species, including IgA, IgM, and IgY. They withstand repetitive prolonged treatment with harsh cleaning agents, and they completely avoid concerns associated with leaching known cytotoxins. They ameliorate the risk of product denaturation somewhat by use of an alternative elution strategy, but the risk is still high in comparison to most other methods. They are inferior to protein G in one respect only: they indiscriminantly capture all antibodies from a sample, specific or not. Protein G copurifies only nonspecific IgG.

thiophilic adsorption versus HIC

Controlled comparisons of thiopyridyl ligands with phenyl HIC ligands have not been published, but phenyl media are known to respond to ethylene glycol in similar fashion.[54,55] Whether phenyl ligands respond to electron donors in the same way remains to be seen. The ability to load these gels at relatively moderate salt concentrations makes them more convenient than weaker hydrophobic ligands. Whether the elevated risk of denaturation justifies the convenience must be assessed on a case-by-case basis.

Dye ligand chromatography

Dye ligands like reactive blue-2 embody complex combinations of charge, hydrophobic and other protein-interactive mechanisms (Figure 8.7). Many of these ligands are strongly anionic, bearing multiple

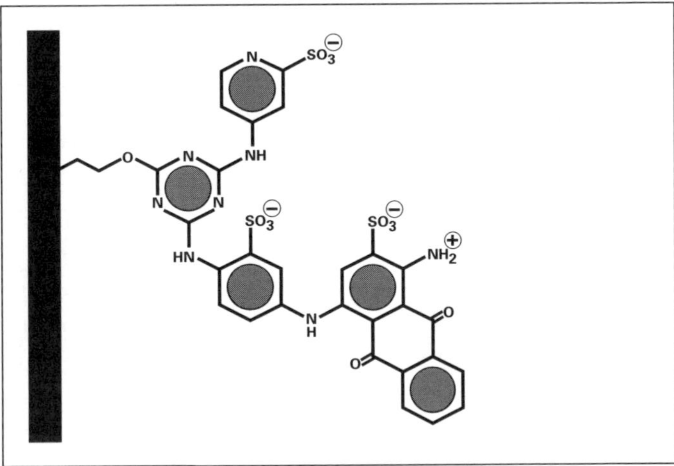

Figure 8.7. Immobilized reactive blue-2. This dye-ligand is also known as cibachron blue F3GA

sulfo groups that endow them with a strong cation exchange functionality. Most monoclonals bind at low pH and can be eluted in a salt gradient. Strongly basic monoclonals will bind under physiological conditions and can be eluted in the same fashion.[56] Remazol yellow-GCL has been reported to have strong affinity for a subfraction of polyclonal IgG but has not been evaluated for monoclonal purification.[57] A cationic yellow dye has been synthesized that may exhibit interesting variations of anion exchange.[58]

hybrid gels

Reactive blue-2, also referred to Cibachron blue F3GA, has been co-immobilized with DEAE and applied successfully to purification of mouse IgG_1 and IgG_{2a} monoclonals.[59,60] At pH 7.2, the authors developed salt gradient conditions that yielded ~90% pure antibody, free of protease and ribonuclease activity. Mass recovery was estimated at 70–80%. Elution profiles and PAGE gels indicate that the overall selectivity was dominated heavily by the DEAE component of the matrix. Transferrin eluted early in the gradient, followed immediately by IgG, and later by albumin and the bulk of other acidic contaminants.

limitations

In spite of the apparent value of this "one-step" application, dye ligands bear serious limitations. The first is lot-to-lot variability of the dye itself.[61] Second, compound ligands—such as DEAE/blue-2—are more difficult to manufacture consistently than single-ligand gels. Whether the ligands are immobilized sequentially or in tandem, a reproducible ratio is difficult to achieve. Progressive dye leakage compounds the problem. Dye leakage is also a validation burden. The most recent generation of dye ligands are reported to have overcome this problem.[62-64] However, selectivities customized for monoclonal purification have yet to emerge.

ABx

Where complex dye ligands aspire to mimicry of biological structures, ABx represents an alternative philosophical approach. It hybridizes a cation exchanger with mild anion exchange and hydrophobic functionalities.[65] The idea is that simple combinations of physicochemical properties can be developed that complement antibody surface chemistry.[56-58] Antibodies are, after all, among the most basic and hydrophobic proteins in serum. Its perfectly reasonable to expect that a weak-moderate hydrophobic ligand with a strong cation exchange functionality might offer valuable selectivity. In practice, the selectivity of ABx is overwhelmingly dominated by the cation exchange functionality and falls within the range of other commercial cation exchangers.[65-68] Nevertheless, the rationale is sound and represents a promising opportunity for development of new purification tools.

Boronate affinity chromatography

IgM purification on immobilized m-aminophenyl boronate exploits an obscure but interesting mechanism. Immobilized boronates are sometimes referred to as artificial lectins, in reference to their interaction with carbohydrates. Under alkaline conditions, boronate anions form covalent bonds with 1,2 cis-diols (Figure 8.8).[69-76] However, whereas lectins react chiefly with specific terminal carbohydrate moieties, boronates react with cis-diols regardless of the identity or position of the sugar. Bound proteins are eluted by reducing pH or by competition with other cis-diols; principally sugars such as sorbitol.

Unfortunately the specificity of the interaction is influenced heavily by the phenyl groups, which confer varying levels of hydrophobicity depending on ligand density. The technique is hit or miss. When it works

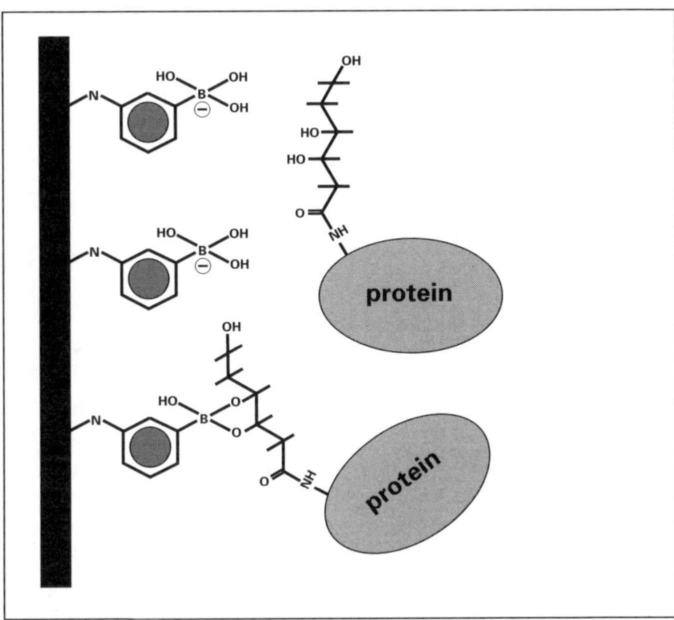

Figure 8.8. Immobilized aminophenyl boronic acid.

at all, antibody purity is 35–65% and mass recovery is often no better. These figures might improve with IgM-optimized adsorbents and systematic development guidelines. However, the method's enigmatic behavior, mediocre performance, and the small commercial opportunity represented by IgM purification seem unlikely to attract gel manufacturers to the cause.

References

1. A. Alpert, 1990, *J. Chromatogr.*, **499** 177
2. J. Palmer, 1975, *Anal. Lett.*, **8** 215
3. F. Rabel et al, 1976, *J. Chromatogr.*, **126** 731
4. M. Rubenstein et al, 1979, *Proc. Nat. Acad. Sci. USA*, **76** 640
5. M. Rubenstein, 1979, *Anal. Biochem.*, **98** 1
6. J. Brewer and L. Soderberg, 1977, in Chromatography of Synthetic and Biological Polymers, Vol. 1, (R. Epton, ed.) p. 285, Ellis Horwood, Chichester
7. S.-C. Yan et al, 1984, *Anal. Biochem.*, **138** 137
8. K. Hellsing, 1968, *J. Chromatogr.*, **36** 170
9. J. Giddings and K. Dahlgren, 1970, *Sep. Sci.*, **5** 717
10. T. Laurent and J. Killander, 1964, *J. Chromatogr.*, **14** 317
11. H. Schachman and M. Lauffer, 1949, *J. Am. Chem. Soc.*, **71** 536
12. J. Lee and L. Lee, 1979, *Biochemistry*, **18** 5518
13. J. Lee and L. Lee, 1981, *J. Biol. Chem.*, **156** 625
14. T. Arakawa, 1985, *Anal. Biochem.*, **144** 267
15. S. Neoh et al, 1986, *J. Immunol. Met.*, **91** 231
16. P. Gagnon et al, 1995, *BioPharm*, **8**(3) 21

17. K. Ingham ,1990, *Met. Enzymol.*, **182** 301
18. D. Atha and K. Ingham, 1986, *J. Biol. Chem.*, **256** 12108
19. L. Lee and J. Lee, 1987, *Biochemistry*, **26** 7813
20. P. Gagnon et al, 1995, A Method for Obtaining Unique Selectivities in Ion Exchange Chromatography by Addition of Organic Polymers to the Mobile Phase, poster, 15th International Symposium on HPLC of Proteins, Peptides, and Polynucleotides, Boston, reprint: <http://www.validated.com/library.html>, manuscript in press, J. Chromatogr., ref. CHROMA 28044
21. R. Kasai et al, 1987, *J. Chromatogr.*, **407** 205
22. K. Knudsen et al, 1992, *Anal. Biochem.*, **201** 170
23. P. Gagnon, 1992, Multiple Mechanisms for Improving Binding Efficiency with Protein A, poster, BioEast '92, Washington D.C., reprint: <http://www.validated.com/library.html>
24. PolyGlycoplex, PolyLC Inc., Columbia, MD USA
25. POROS-ether, PerSeptive Biosystems, Cambridge, MA
26. K. Ingham and T. Busby,1980, *Chem. Eng. Commun.*, **7** 315
27. J.-P. Bouvet et al., 1984, *J. Immunol. Met.*, **60** 299
28. P. Clezardin et al, 1986, *J. Chromatogr.*, **354** 425
29. P. Clezardin et al, 1986, *J. Chromatogr.*, **358** 209
30. C. Middaugh and G. Litman, 1977, *J. Biol. Chem.*, **252** 8002
31. S. Timasheff and G. Fasman, 1969, Structure and Stability of Biological Macromolecules, Marcle Dekker, New York
32. W. Jencks, 1969, Catalysis in Chemistry and Enzymology, p.323, McGraw-Hill, New York
33. C. Tanford, 1968, *Adv. Protein Chem.*, **23** 121
34. J. Porath and N. Ui, 1964, *Biochim. Biophys. Acta*, **90** 324
35. E. Kabat and M. Mayer, 1966, Experimental Immunochemistry, 2nd ed., p.264, Charles C. Thomas Publisher, Springfield
36. U.-B. Hansson and E. Nilsson, 1973, *J. Immunol. Met.*, **2** 221
37. J. Porath et al, 1985, *FEBS Lett.*, **185** 306
38. J. Porath, 1986, *J. Chromatogr.*, **376** 331
39. T. Hutchins and J. Porath, 1986, *Anal. Biochem.*, **159** 217
40. J. Porath, 1987, *Biopolymers*, **26** S193
41. M. Belew et al, 1987, *J. Immunol. Met.*, **102** 173
42. J. Porath and M. Belew, 1987, *Trends Biotechnol.*, **5** 225
43. S. Oscarsson and J. Porath, 1989, *Anal. Biochem.*, **176** 330
44. B. Nopper et al, 1989, *Anal. Biochem.*, **180** 66
45. A. Lihme and P. Heegaard, 1991, *Anal. Biochem.*, **192** 64
46. Prosep-Thiosorb and Prosep-ThioSorb-M, Bioprocessing Ltd., Consett UK
47. S. Timasheff and H. Inoue, 1968, *Biochemistry*, **7** 2501
48. T. Ngo and N. Khatter, 1992, *J. Chromatogr.*, **597** 101
49. T. Ngo and N. Khatter, 1991, *Appl. Biochem. Biotechnol.*, **30** 111

50. N. Khatter et al, 1991, *Am. Lab.*, **May** 32LL
51. D. Narinesingh et al, 1991, *Anal. Lett.*, **24** 2005
52. T. Ngo and N. Khatter, 1990, *J. Chromatogr.*, **510** 281
53. T. Ngo, 1991, US Patent 4,981,961
54. P. Gagnon and E. Grund, 1996, *BioPharm*, **9**(5) 54
55. —1990, Hydrophobic Interaction Chromatography, Principles and Methods, Pharmacia Biotech, Upssala,
56. S. Fulton, 1989, The Art of Antibody Purification, Amicon, Danvers, MA USA
57. P. Byfield, 1989, in Protein and Dye Interactions: Development and Applications, (M. Vijayala and O. Bertrand, eds.) p. 244, Elsevier, Barking
58. Y. Clonis et al, 1987, *Biotechnol. Bioeng.*, **30** 621
59. C. Bruck et al, 1982, *J. Immunol. Met.*, **53** 313
60. C. Bruck et al, 1986, *Met. Enzymol.*, **1221** 587
61. R. Scopes, 1987, *Anal. Biochem.*, **165** 235
62. K. Jones, 1991, *LC-GC*, **4**(9) 32
63. S. Santambien et al, 1995, *J. Chromatogr.*, **664** 241
64. —1995, Toxicity Evaluations of Early Textile Dyes and Current Biomimetic Derivatives, Prometic Biosciences, Burtonsville, MD USA
65. D. Nau, 1989, *BioChromatography*, **4** 4
66. D. Nau, 1986, *BioChromatography*, **1** 82
67. D. Nau, 1897, in Commercial Production of Monoclonal Antibodies, (S. Seaver, ed.) p. 247, Marcel Dekker, New York
68. A. Ross et al, 1987, *J. Immunol. Met.*, **102** 227
69. J. Boeseken, 1949, *Adv. Carbohydr. Chem.*, **4** 189
70. A. Foster, 1957, *Adv. Carbohydr. Chem.*, **12** 81
71. S. Wetzman et al, 1979, *Anal. Biochem.*, **97** 438
72. S. Narasimhan, et al, 1980, *J. Biol. Chem.*, **255** 4876
73. R. Rothman and L. Warren, 1988, *Biochim. Biophys. Acta*, **955** 143
74. T. Reid and D. Gisch, 1988, *BioChromatography*, **3** 201
75. J. Mazzeo and I. Krull, 1989, *BioChromatography*, **4** 124
76. GlycoGel-B, Pierce Chemical Company, Rockford, IL

Chapter 9

Protein A Affinity Chromatography

"When a thing ceases to be an object of controversy, it ceases to be a thing of interest."
—William Hazlitt

On July 10, 1972, manuscripts from 2 independent research groups, both describing affinity purification of IgG on immobilized protein A, arrived at the editorial offices of 2 different journals.[1,2] The rest, as they say, is history. With the emergence of monoclonals, protein A quickly became the purification method of choice among researchers and soon evolved to become a popular method in commercial circles.

Mechanisms

Protein A (Cowan strain I) is a cell wall component of *Staphylococcus aureas*. It consists of a single polypeptide chain in the form of an elongate nodular cylinder.[3,4] The C-terminus begins with a cell wall/membrane-associated region, proceeding into a linear series of 5 highly homologous antibody-binding domains (Figure 9.1).[5-9] These domains are roughly cylindrical (26 x 16Å) with individual molecular weights of ~6.6kD.[10,11] Each has approximately equivalent ability for antibody binding.

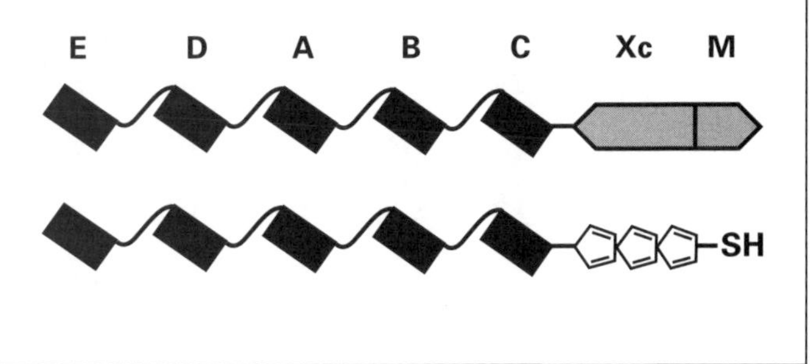

Figure 9.1. Domain structure of wild-type and recombinant protein A. Black rectangles indicate IgG-binding domains. Xc and M indicate the transmembrane and trans cell wall domains of Cowan strain. Pentagons indicate histidyl residues on recombinant protein A. The sulfhydryl indicates a terminal cysteine for gel anchorage

The flexible random coil sequences that link the domains provide a fair degree of steric mobility but are very susceptible to proteolytic cleavage.[10-13] The binding domains themselves are protease resistant. The molecular weight of the intact molecule is ~54kD.[14] This is calculated from the gene and represents the mature protein after cleavage of a membrane-transport signal sequence. Molecular weight estimates from PAGE average 42–45kD, indicating resistance to unfolding in SDS.[15] Lower estimates reflect inadequate control of proteolysis during purification.

recombinants

Recombinant protein As are mostly derived from secreted extracelluar variants that lack the cell-wall associated region.[16,17] Compared with Cowan strain I, there are minor differences in the amino acid sequences of the IgG binding domains, but antibody reactivity is indistinguishable. The C-terminus is often altered to facilitate purification of the protein A itself, for example by the insertion of polyhistidyl sequences to enable purification by immobilized metal adsorption chromatography (IMAC). Constructs may also incorporate features to support directional coupling of the ligand to solid phase supports. One such recombinant incorporates a C-terminal cysteine.[18] Thioether coupling allows the ligand to extend farther into the mobile phase space than would be possible for a laterally immobilized ligand. This mimics the orientation believed to exist on the bacterial cell wall and improves antibody binding.[11,18] Other recombinants incorporate C-terminal polylysyl sequences for the same purpose.

synthetic analogs

The main advantage of recombinant protein A is that it precludes copurification of toxic contaminants from a pathological bacterial source. A recently introduced synthetic protein A-mimetic ligand suspends all concerns pertaining to biological origin.[19] The one chromatography product currently available is based on a single IgG binding domain.[20] Immobilized digest-derived single domains from natural protein A exhibit better IgG binding capability than the intact molecule.[21] Single domains also lack the protease-labile interdomain sequences. These points likely con-

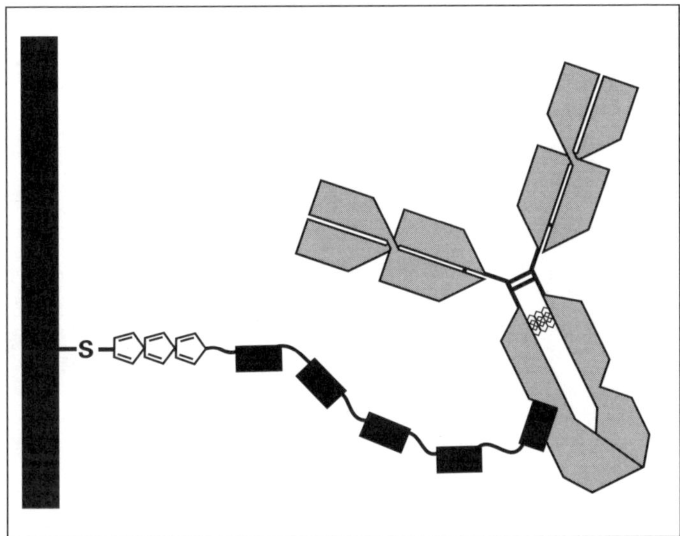

Figure 9.2. Protein A binding to IgG-Fc. See text for details

tribute to the high IgG binding and low leaching levels observed with the synthetic ligand.[20,22] Unlike most recombinants, composition of the mimetic ligand's IgG-binding site has been altered. Preliminary data indicate that its selectivity is similar to natural and recombinant forms, but detailed studies are required to determine the extent of the similarity.

primary binding site

The primary binding site for protein A is on the Fc portion of IgG at the junctures of the Cγ2 and Cγ3 domains (Figure 9.2).[10,23] Experimental data indicate that induced fit occurs, explaining the harsh conditions required for elution. X-ray crystallographic measurements obtained at 2.8Å resolution show that the Cγ3 domains are unaffected. The Cγ2 domains are displaced longitudinally toward the Cγ3 domains; locked into place by association with single protein A domains. Besides altering the conformation of the binding sites, this destabilizes the distal third of the Cγ2 domains, which causes partial rotation and destabilization of the carbohydrate region between the Cγ2 domains. Aside from the consequences of harsh elution conditions on the antigen-binding capabilities of protein A-purified antibodies, these secondary effects raise concerns about altered antibody effector functions and increased susceptibility to proteolysis.[24,25]

Figure 9.3. Binding of Mouse IgG_1 to protein A as a function of salt concentration. A: sodium sulfate. B: sodium chloride. All experiments were run in 0.05M boric acid, pH 9.0. Binding values expressed as % of total antibody applied to the column. Note the reduction in binding above 1.0M sodium sulfate. This is from precipitation of the antibody.

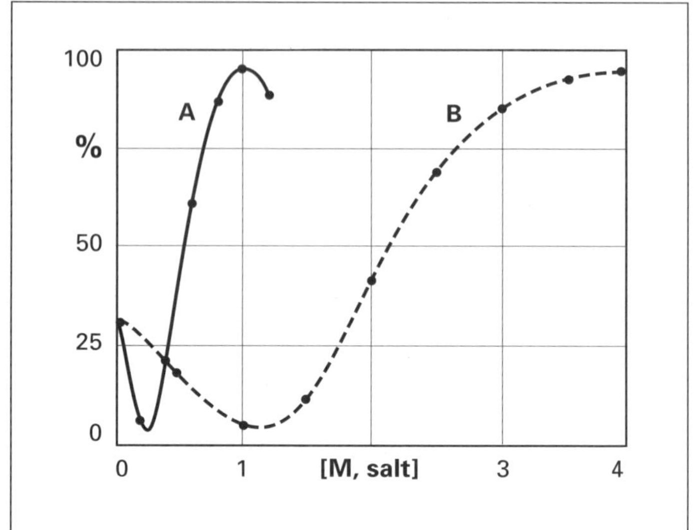

binding energy

The majority of the binding energy comes from hydrophobic interactions.[10,23] Water is displaced from an area of 1234Å^2. The association is stabilized by 4 hydrogen bonds but there are no ionic bonds. As expected for a dominantly hydrophobic association, binding is strongly responsive to salts.[26-29] Figure 9.3 illustrates binding efficiency of a mouse IgG_1 to protein A as a function of sodium sulfate and sodium chloride concentration. The responses follow the classic salting-in salting-out curve.

variations in effectivity of different binding salts

Observations indicate that the molar effectivity of various salts for promoting protein A binding follows Hofmeister rankings, as with hydrophobic interaction chromatography (HIC).[30,31] In spite of many salts being able to enhance protein A-binding to IgG, relatively few are well qualified for the task. Sodium chloride at 3.0–5.0M is often used, as is sodium sulfate up to 1.0M.[29] Potassium phosphate is effective so long as its strong buffer capacity is not a problem.[31] Ammonium sulfate is effective at neutral pH but strongly basic free ammonia liberated above pH 7.5 denatures proteins.[31-34] Volatilization of ammonia also contributes to buffer pH instability and can be a significant health hazard at large process scales.[31,33] Consistent with their behavior in HIC, chaotropic salts

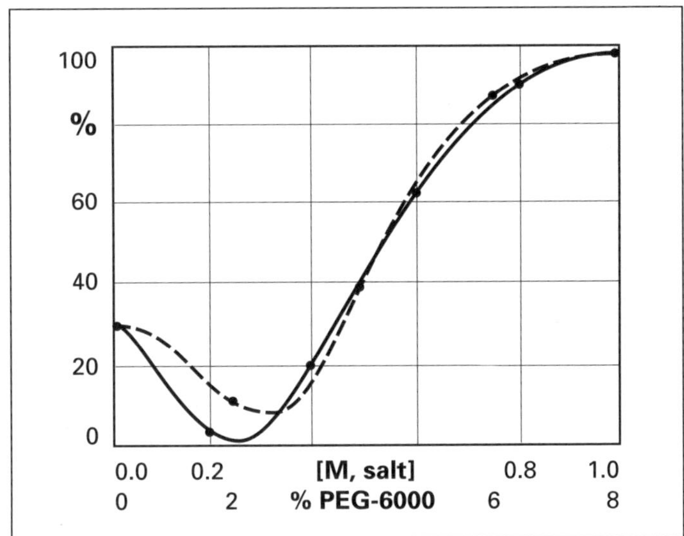

Figure 9.4. Binding of Mouse IgG_1 as a function of PEG-6000 and sodium sulfate concentration. The solid curve represents sodium sulfate. The dashed curve represents PEG-6000. All experiments were conducted in 0.05M boric acid, pH 9.0. Binding values expressed as percent of total antibody applied to the column.

like magnesium chloride, potassium iodide, and sodium isothiocyanate have an eluting effect.[29,31,35,36]

binding enhancement with polyethylene glycol

Polyethylene glycol (PEG) produces a response similar to precipitating salts (Figure 9.4).[37] This is attributed to a related mechanism. Like precipitating salts, PEG is excluded from protein surfaces, leaving them preferentially hydrated.[38-43] With increasing PEG concentration in the bulk solution, the association between protein A and IgG is enhanced by entropically driven sharing of their hydration shells.[38] Although interesting from a mechanistic perspective, the high viscosity of PEG solutions makes this approach impractical for most preparative applications.

binding enhancement with glycine

Other enhancing additives include glycine, which is often added to protein A binding buffers in high molar concentrations. However, its effect is modest in comparison to its price: 1.0M sodium sulfate plus 2.0M glycine has about the same effectivity as 1.1M sodium sulfate.[31,37] Glycine also enhances antibody stability.[44-46]. Its first pK of 2.35 is too low to qualify it as an elution buffer and its second pK of 9.76 is too high for a binding buffer.[47]

effects of hydrophobic competitors

Elution of protein A columns with weakly hydrophobic competitors like glycyl-tyrosine and ethylene glycol has been only partially successful.[48]

Figure 9.5. Binding of Mouse IgG_1 as a function of pH. Binding expressed as % of total IgG applied to the column. All experiments conducted in 1.0M sodium sulfate.

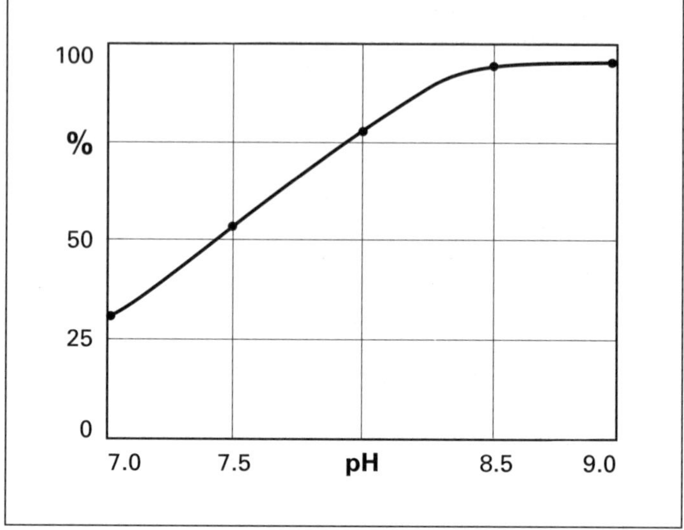

Antibodies can be dissociated with moderate concentrations of stronger organic solvents, such as methanol (20–30% v:v), but the probability of permanent antibody denaturation disqualifies them from preparative applications. Elution is most commonly achieved by reducing pH. This also involves risk of denaturation, but less than with strong organic solvents. Electrophoretic elution has not proven practical.[49,50]

pH response

The responsiveness of protein A binding and elution to pH is a function of hydrophobicity (Figure 9.5). Studies reveal a highly conserved histidyl residue in the center of the protein A-binding site of IgG (Table 9.1).[51-53] This residue aligns facing a complementary and similarly conserved histidyl residue on protein A.[5-9,16] At alkaline pH, these residues are uncharged and there are no restrictions on interfacial contact. The hydrophobic character of the uncharged imidazolium rings contributes to net hydrophobicity at the interface, strengthening the association. At low pH, the histidyls are fully charged, minimally hydrophobic, and mutually repellent.

Monoclonals that require high pH for binding can be eluted by modest pH reductions. Antibodies that bind strongly under physiological conditions are less responsive. Antibodies that lack the critical histidyl

Table 9.1. Protein A-contact residues and hydrophobic indices for mammalian IgGs. "HG" indicates Human IgG; "HM" indicates murine. The subscript indicates class. The "key" histidyl residues are indicated in bold. M: methionine. T: threonine. I: isoleucine. S: serine. V: valine. L: leucine. H: histidine. Q: glutamine. N: asparagine. B: glutamic acid. D: aspartic acid. Y: tyrosine. "H. index" is the cumulative transfer coefficent of the individual residues, in thousands (ref. 52). Sequence data from refs. 51,53. Note the level of conservation and the high concentration of histidine residues.

Antibody:	**HG_1**	**HG_2**	**HG_3**	**HG_4**	**MG_1**	**MG_{2a}**
contact residues:	M	M	M	M	T	M
	I	I	I	I	I	I
	S	S	S	S	T	S
	V	V	V	V	I	I
	L	V	L	L	M	Q
	H	**H**	**H**	**H**	**H**	**H**
	Q	Q	Q	Q	Q	Q
	N	D	B	D	D	D
	H	H	H	H	H	H
	N	N	N	N	N	N
	H	H	H	H	H	H
	Y	Y	Y	Y	H	H
H. index:	11.1	10.8	11.1	11.1	10.0	8.9

residue don't bind at all. These mainly include human IgG_3s of the Caucasian allotype G3m (s^-t^-).[23,54-58]

temperature effects

The only data not directly supportive of protein A:IgG interactions being principally hydrophobic come from a study showing that binding is improved at cold temperatures.[59] This is contrary to experimental results showing weaker reduced-temperature binding of IgG on HIC media.[31,60,61] The authors reconciled the variance by suggesting enhanced hydrophobic contact occurred as a result of conformational changes restricted to the binding site.[59]

non-IgG binding

Evidence exists for protein A binding to the Fc regions of IgAs and IgMs, some of which respond to elevated pH and salt concentration in the same way as IgGs.[62-69] In addition, an alternative binding pathway has been described whereby protein A binds the constant region of Fab from all antibody classes.[63-66,70-81] Antibodies with kappa light chains bind more strongly than those bearing lambda light chains.[82] Unlike the Fc pathway, Fab binding appears to have a significant ionic component. This is suggested by diminishing capacity of some IgA monoclonals with increasing sodium chloride concentration up to 2.0M.[66] All the same, that binding survives above

Table 9.2. Protein A affinity for mammalian IgG from different species and subclasses. Data from compilations in references 17, 29, and 53. Information should be regarded as approximate since it was obtained under a range of salt concentrations on many different chromatography supports. G. Pig indicates Guinea Pig.

IgG species /subclass	affinity	binding pH	elution pH
Human 1	high	7.0–7.5	2.5–4.5
Human 2	high	7.0–7.5	2.5–4.5
Human 3	none/mod.–high	7.0–7.5	3.0–7.0
Human 4	high	7.0–7.5	2.5–4.5
Mouse 1	low	8.5–9.0	6.0–7.0
Mouse 2a	low–mod.	8.0–9.0	4.5–5.5
Mouse 2b	mod.–high	8.0–9.0	3.5–4.5
Mouse 3	low–high	8.0–9.0	4.0–7.5
Rat 1	low	8.0–9.0	6.0–8.0
Rat 2a	none–low	9.0	7.5–9.0
Rat 2b	low	8.0–9.0	7.0–8.0
Rat 2c	low–high	8.0–9.0	3.0–7.0
G. pig 1	high	7.5–9.0	4.0–5.0
G. pig 2	high	7.5–8.0	3.0–4.5
Rabbit	low-high	7.5	3.0–7.0
Bovine 1	low	7.0–8.0	6.0–7.0
Bovine 2	high	7.0–7.5	3.0–4.0

0.50M indicates a hydrophobic contribution, and it's likely that high concentrations of precipitating salts enhance Fab binding as they do Fc binding. The 2 binding pathways act cooperatively in any case.[83]

variations in affinity for different antibodies

Variations in affinity for IgGs from different species and subclasses are well known (Table 9.2).[17,29,53,84-100] Comparison of Tables 9.1 and 9.2 shows some correlation between affinity and net hydrophobicity of IgG/protein-A contact residues but also some deviations. Stronger binding by human IgG is consistent with the net hydrophobicity of its contact residues, but the less hydrophobic mouse IgG_{2a} binds more strongly than mouse IgG_1. This is a reminder that although the binding energy derives predominantly from hydrophobic interactions, highly specific alignment is required to support it. The process value of this point is that modification of buffer conditions to enhance binding of low affinity antibodies is unlikely to promote binding by nonimmunoglobulin contaminants.

Attributes

Protein A's most compelling feature is its simplicity: load it, wash it, elute it. Purification can be conducted quickly and easily with the most rudimentary or the most sophisticated instrumentation. Using a hand held syringe setup with commercial binding buffers you can obtain 5-20 mg of 90% pure antibody in under 15 minutes. With the recent generation of high flow chromatography media on an HPLC you can achieve the same result in under 3 minutes.[101] With appropriate binding buffer formulation, most mammalian IgGs adsorb well enough to support efficient purification.

applicability

Human and guinea pig monoclonals bind sufficiently well under physiological conditions that no pre-equilibration is required. Mouse IgG_{2a}, IgG_{2b}, IgG_3, and rat IgG_{2c} bind under these conditions, but capacity is improved by elevating pH and/or salt concentration. Rat IgG_1, IgG_{2b} and mouse IgG_1 require high pH in combination with salt concentrations approaching antibody solubility limits to achieve adequate binding. Even then, capacities are substantially lower than they are for strong-binding antibodies. Binding of rat IgG_{2a} is marginal under any conditions.

purification performance

Protein A is capable of providing total IgG purity as high as 95% in a single step, but specific monoclonal purity is rarely this good. Protein A co-concentrates nonspecific antibodies. Ascites average about 1 mg of host polyclonal IgG per mL.[76] This means that specific monoclonal purity will seldom be as good as 90% and may be no better than 65%. Even these figures neglect protein A binding up to 30% of polyclonal IgA and up to 65% of polyclonal IgM.[77,102,103]

Antibodies produced in cell culture often fare no better. Even though serum supplementation levels are often low, antibody production levels are proportionally even lower. "Fetal" bovine serum may contain up to 1 mg/mL of IgG, although the usual range is 40–400 µg.[104-110] A monoclonal that produces poorly in a heavily supplemented medium may represent <50% of the purified antibody.[111]

Contamination by nonspecific antibodies can be reduced significantly with gradient elution, but not

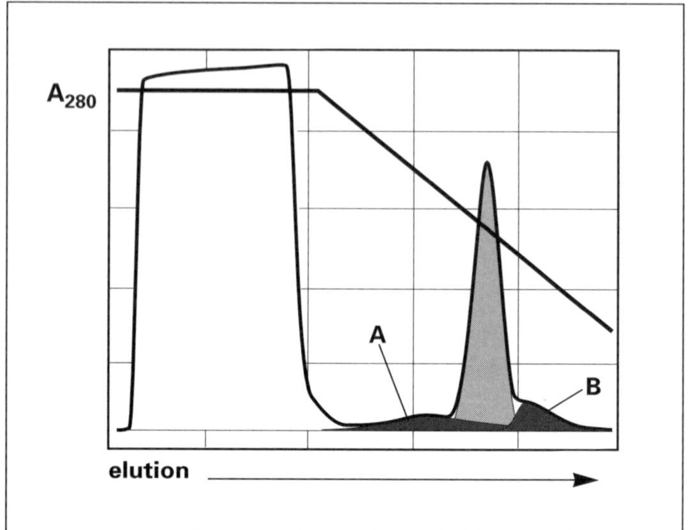

Figure 9.6. Partial fractionation of bovine IgG (A) and antibody aggregates (B) by linear pH gradient elution. The main elution peak is mouse IgG_{2a}. Cell culture supernatant applied in 0.02M boric acid, 0.02M sodium phosphate, 0.02M sodium citrate, 1.0M sodium sulfate, pH 9.0. Eluted in a 10CV linear gradient to 0.02M sodium phosphate, 0.02M sodium citrate, 0.5M sodium chloride, pH 3.0.

eliminated (Figure 9.6). Additional methods are required, such as ion exchange chromatography (IEC), which provides ~10 times the nonspecific antibody removal ability of protein A.[110]

purification of non-IgG monoclonals

The high percentage of non-IgGs that bind protein A make it a candidate for their purification. However, optimization is complicated by the unknowable relative contributions of Fc and Fab binding. Binding of some non-IgGs is enhanced at elevated pH, but to varying degrees.[66-69] Some bind better in high salt.[67-69] Some are salt-independent.[66] Some lose capacity.[66] Binding capacities are often lower for nonIgGs, especially when they have to compete with IgG for binding substrate. If you plan to screen protein A for non-IgG purification, remove the IgG first by immobilized metal affinity chromatography (IMAC) or affinity chromatography on immobilized protein G. Note that even though many IgMs bind, the harsh elution are often detrimental to their performance.

aggregate binding

Antibody aggregates bind more strongly than native antibody, possibly due to a larger number of interactions with the column. Linear pH gradients are able to partially resolve them from native antibody (Figure 9.6).[29] Removal may exceed 80%. Step gradients are less effective. Antibody fragments bind more weakly than intact parent antibody.[29] This probably reflects the loss of cooperativity between Fc and Fab binding.

Speculation has been raised that protein A may selectively enrich or eliminate post-transcriptional or post-translational variants from purified monoclonal preparations.[17] This is an issue with charge-based separations like IEC and hydroxyapatite (HAC) but not with protein A. Protein A binding is unaffected by variations in glycosylation—even complete deglycosylation.[112,113] If protein A-purified monoclonals exhibit performance aberrations suggestive of selective glycoform fractionation, they more likely reflect conformational modification induced by harsh elution conditions or secondary effects of leached protein A.

non-immunoglobulin contaminants

Protein A specifically binds and copurifies both α_2-macroglobulin and kinogen along with antibodies.[114] Elevated salt concentration has been reported to co-enhance nonspecific binding of other nonimmunoglobulin contaminants.[115] This probably relates to nonspecific interactions with an uninterrupted sequence of 21 hydrophobic residues in the transmembrane region of wild-type protein A.[116] Nonspecific contaminant binding in high salt is negligible in practice and should be eliminated altogether by use of recombinants that lack the hydrophobic sequence.[16,17] To the extent that such binding does occur, the contaminants are easily removable with simple adjustments to the wash or elution conditions.

DNA removal

Protein A reduces DNA contamination by up to 4.5 logs.[117-119] Its ability to support antibody binding at high salt concentrations is especially valuable because it confers the ability to dissociate ionic complexes that may exist between DNA and the product. Complex-dissociation significantly improves both product-capture efficiency and DNA clearance.[120] High salt loading is therefore good process insurance even if not required for product binding. If you don't want high salt in the eluted antibody fraction, simply omit it before elution. Even weakly-bound monoclonals will usually tolerate this. Chapters 3, 4, and 6 contain more detailed discussion of DNA complexation.

Table 9.3. Physicochemical resistance of protein A. Refer to references in the text for details and time limits for exposure.

Agent
1.0M sodium hydroxide
12mM hydrochloric acid
0.2M phosphoric acid
5.8M acetic acid
0.5M acetic acid, 60% ethanol
70% ethanol
1.0M potassium iodide
4.0M magnesium chloride
2.0M sodium isothiocyanate
6.0M guanidine-HCl
8.0M urea
80°C, for 10 minutes

virus removal

Protein A removes up to 6 logs of virus.[118,119, 121,122] Low pH elution may enhance viral inactivation but approach this carefully. Prolonged exposure to virucidal pH may inactivate the antibody as well.

endotoxin removal

Protein A purification can reduce endotoxin contamination by 1-4 logs.[119,121-123] The inability of endotoxins to bind HIC media at high salt concentrations suggests that high salt levels are unlikely to detract from protein A's endotoxin removal ability. It is important to appreciate that poorly maintained columns can become endotoxin generators no matter what the inherent removal capability of the chemistry. Good column hygiene is essential.

mass recovery

Mass recovery of specific IgG from protein A is typically 80–90% under appropriate conditions. Lower recoveries warn of inadequate binding or elution conditions, overaggressive peak cutting, or column overloading. Bear in mind that an alarmingly low recovery of total IgG may mask outstanding recovery of specific antibody. This emphasizes the importance of being able to discriminate specific from nonspecific antibody throughout method development.

high capture efficiency

One of the chief advantages of affinity chromatography is the ability to load columns without sample pre-equilibration. However, protein A fulfills this ideal only with human IgG. Low affinity for mouse and rat

Table 9.4. Cleaning/sanitization of protein A columns. Wash 1 is directed against DNA. Wash 2 and 3 are directed against lipid and endotoxin. Wash 4 removes protein contaminants and detergent. The publication from which this protocol is derived uses the DNAase in conjunction with sample loading to prevent DNA fouling, rather than remove it after the fact. This requires subsequent validation of DNAase removal.

Wash 1: 4–6 hours at 21°C with 0.02M sodium phosphate, 2mM magnesium chloride, 13000 units/L DNAase, pH 8.3, 20cm/hr.

Wash 2: 2 hours with 1mN sodium hydroxide, 2M sodium chloride, 20% ethanol, 20cm/hr.

Wash 3: 1.5 hours with 0.02M sodium phosphate, 5% sodium lauryl sarcosinate, 0.1M sodium chloride, 20mM EDTA, pH 7.0, 20cm/hr.

Wash 4: with 50mM acetic acid, 20% ethanol, 2CV at 100cm/hr.

Wash 5 and store: 20% ethanol, 2CV at 100cm/hr.

antibodies necessitates sample pre-equilibration with enhanced binding buffer formulations.

resistance to harsh sanitizing conditions

Protein A is remarkably resistant to physicochemical stress (Table 9.3).[3,15,18,35,83,101,124,125] This should not be understood to indicate that all support media and immobilization chemistries are equally tolerant of this range of conditions. They aren't. Nonetheless, protein A's physicochemical robustness is an important factor in its continued domination of biological affinity methods for antibody purification. Table 9.4 offers a cleaning/sanitization protocol adapted from reference 126. This is a very broad spectrum protocol from which you can adapt the steps most germane to your particular needs. Be sure to confirm chemical compatibility in advance with the gel manufacturer.

sample preparation

The ability of a support to withstand harsh treatment is not a substitute for proper sample preparation.[127] Because of its ability to capture antibody efficiently from dilute crude sample streams, and in spite of its high procurement expense, protein A is often used as the initial—or only—process step. Some users precede the protein A support with a precolumn packed with buffer exchange media. The precolumn is taken off-line after washing. This practice removes gross cell debris, clots and precipitates but has little effect on lipids, endotoxin or DNA. Even so, it typically extends column life by a factor of 10 over sample preparation by membrane filtration.[128]

Figure 9.7. Loss of titer and amplification of nonspecific interference from protein A purification. A: protein A-purified Mouse IgG_1, eluted at pH 5.0 and titrated immediately to pH 7.0. B: nonaffinity-purified control, prepared by IEC and HIC. Curves like the protein A-purified curve are often referred to as sliders, in reference to their indefinite titers. Conversion to nondenaturing physicochemical methods usually normalizes performance.

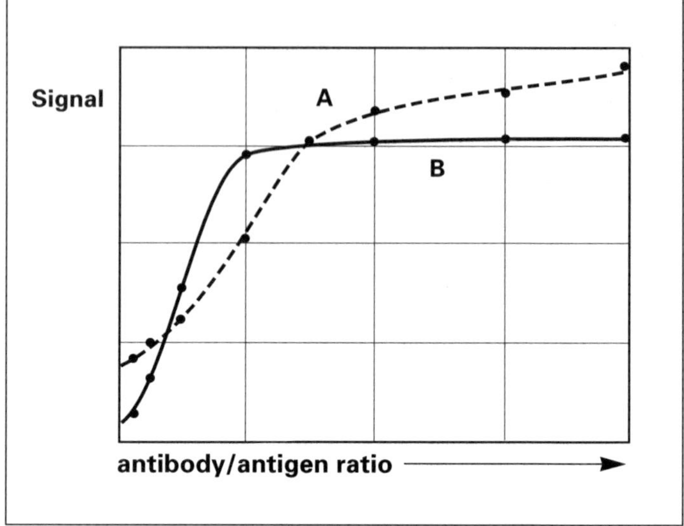

A more effective variation is to replace the buffer exchange media with a high-throughput macroporous anion exchanger. Along with the materials removed by the former, this removes cell micro-debris, endotoxins, DNA, lipoproteins, phospholipids, phenol red and other anionic foulants (appendix II). Expect this treatment to extend media life to well over 100 runs. With media dedicated to purification of a single product from low-protease feedstreams, column life may rival nonbiological media.

Limitations

The most widely held concern with protein A purification is antibody denaturation. It can be manifested as aggregation, fragmentation, loss of immunoreactivity, secondary immunological side-effects in therapy, elevated nonspecific noise in assays, and reduced product stability (Figure 9.7).[17,29,53,88,89,114,129,130]

elution conditions

Most aspects of denaturation are attributed to harsh elution conditions. Low pH is known to permanently alter IgG conformation, reduce titer, and confound effector functions.[131-133] However, frequent inferior performance by mouse IgG_1 monoclonals eluted under mild conditions, in head-to-head comparisons with nonaffinity-purified antibodies, suggests a more fundamental cause. As noted previously, protein A:IgG binding destabilizes important regions of the antibody,

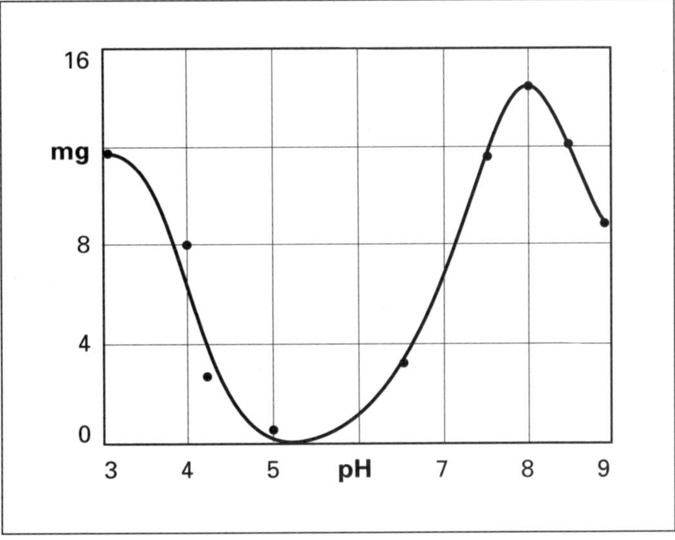

Figure 9.8. Proteolysis of IgG by co-purified protease as a function of pH. Purified IgG was put into dialysis and the buffer monitored for peptides. Values expressed as mg peptide released per g IgG. No proteases were added. Data replotted from reference 142.

and may have permanent effects on antibody performance. The disrupted Cγ2 domain is the residence of several important receptors.[53] The disoriented carbohydrate moieties are involved in a wide variety of secondary functions, including pharmacokinetics and stability.[53,134-140]

proteolysis

Fragmentation is a concern in two contexts. The alkaline conditions required for binding most nonhuman antibodies favor activity by the trypsin proteases (Figure 9.8).[141,142] This makes it important to defer pH equilibration of the sample until immediately before application. The acidic conditions required for elution favor activity by the aspartic and cathepsin proteases.[134] This makes it important to restore pH to neutrality soon after elution.

moderating elution conditions

Minimizing denaturation is necessarily a focal point of method development regardless of the mechanisms involved. The first priority is to moderate the harshness of elution conditions. Linear pH gradient elution is a good tool for determination of minimum elution requirements, as well as for identifying the conditions that provide the best discrimination from contaminating antibodies, fragments, and aggregates. Antibodies that elute above pH 4.5 seldom experience gross denaturation problems.

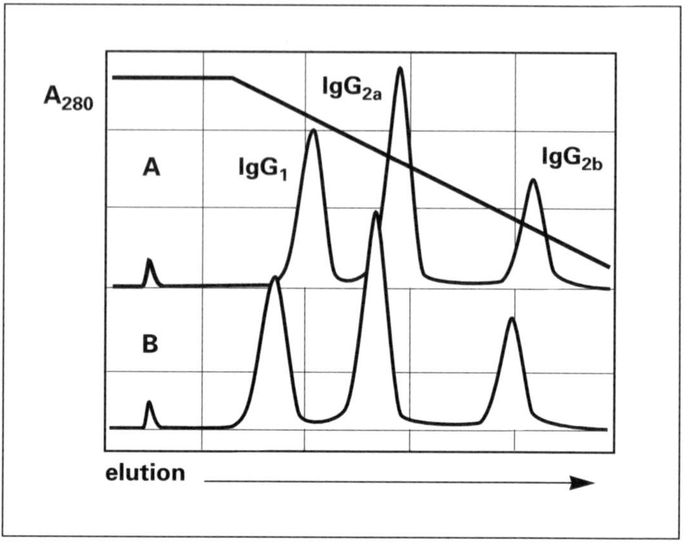

Figure 9.9. Moderation of elution pH by the use of additives. The gradient interval is from pH 9.0 to 3.0. Sample: a cocktail of purified mouse IgG_1, IgG_{2a}, and IgG_{2b}. The buffers for the upper profile contain 0.1M sodium chloride. The lower profile is the same except substituting 0.2M sodium sulfate for the sodium chloride. The effect of salts and other additives can also be assessed in experiments such as illustrated in Figure 9.4, but the process is much more laborious.

nondenaturing elution-enhancing additives

Antibodies that require lower pH often benefit from the use of elution enhancing additives. At least 0.1M sodium chloride should be included to counteract the tendency of antibodies to precipitate at low ionic strength. Higher salt concentrations, up to ~1.0M sodium chloride or 0.2M sodium sulfate, moderate the pH required for elution. To determine the salt concentration that supports the most moderate elution pH for your monoclonal, do a series of experiments at different salt concentrations, maintaining level concentration across a linear pH gradient elution (Figure 9.9).

Ethylene glycol moderates elution pH by weakening hydrophobic interactions.[31,48] It is nondenaturing for some proteins at concentrations up to 30%, and for others at up to 60%.[31,48,143,144] This recommends keeping the concentration towards the lower limit. Another reason to do so is that ethylene glycol increases viscosity and may require compensatory reduction of flow rate. Ethylene glycol is hygroscopic, absorbing up to twice its weight in water.[145] Variations in water content can cause reproducibility problems, so make sure that manufacturing SOPS specify appropriate material handling and storage precautions.

Another benign and useful additive is 1.0–2.0M urea. It is an effective hydrogen donor/acceptor, capable of outcompeting the hydrogen bonds that stabilize the interaction between protein A and IgG.[10, 146,147] Its denaturative influence is negligible in this concentration range.[143,147,148] As with salt and ethylene glycol, evaluate its effects in the context of linear pH gradients.

post-elution re-equilibration

Antibody re-equilibration requires as much attention as elution conditions. Opinions are divided on the best approach. One camp advocates titration immediately following elution. The other camp posits that re-equilibration should be gradual and favors the idea of overnight dialysis to restore the antibody to native form. The rationale is that sudden re-equilibration may snap the antibody into a stable but non-native conformation. Besides the obvious logistical limitations of large-scale dialysis, a concern with this approach is the extended exposure of the antibody to low pH.

Buffer exchange chromatography into a neutral buffer can be effective but it involves a separate process step, with all the accompanying setup and documentation. There may also be a significant time lag between protein A elution and antibody re-equilibration.

Collecting the elution directly into a buffer concentrate is a good compromise. These concentrates should be pretitrated to pH 7.5 and formulated at a high enough concentration so that when they are diluted by the intended fraction volume the final pH is no lower than 6.0. This treatment also reduces the risk of secondary proteolysis by copurified enzymes. Neutralizing-buffer concentration will vary according to the composition and pH of the elution buffer.

Resist the temptation to use higher pH concentrates, such as untitrated Tris (pH~10.3). The abrupt transition of the antibody from low to high pH is a legitimate concern, and the early part of any given fraction may suffer as much from high pH exposure as it would from low pH. Direct addition of base should be avoided for similar reasons. Unless specifications are developed whereby a defined volume of base at a particular concentration is added to a defined volume of sample, titration will not be consistent. Even if relative proportions are defined, uncontrolled addition results in local excesses that denature antibody at the interface. Overshooting pH may compound the problem by requiring backtitration with acid.

leaching

Protein A columns are notorious for leaching. Published evaluations document up to 630ng of leached protein A per mg of IgG.[149-154] Assuming the leachate is proteolysed into single domains, this corresponds to a molar ratio of ~1:70, meaning that up to ~1.5% of the IgG may be contaminated. Higher and lower leachate values are encountered in the lab, depending on the feedstream composition, sample loading conditions, and the column itself.

leaching mechanisms

Leaching occurs by 3 different pathways: breakdown of the support matrix, breakdown of the immobilization linkage, and proteolytic cleavage of the interdomain sequences of protein A. Numerous assay systems have been developed for monitoring leachate.[149-155] The latest generation are capable of detecting as little as 25pg per mg of antibody.[155]

leaching by proteolysis

Leaching is generally highest during a column's first few runs. This has been attributed to unbound protein A left over from the immobilization process.[149-152] However this hypothesis overlooks the fact that leakage is negligible in the absence of IgG.[152,153] The occurrence of leakage with even commercially purified polyclonal IgG preparations probably reflects their ubiquitous contamination with proteases.[142,156-165] These particularly include plasmin and kallikrein, both members of the trypsin family proven so effective for digestion of protein A.[12,13]

This suggests that higher leaching in a column's first few runs reflect higher steric accessibility of the protein A. After the more accessible populations are cleaved off, leaching diminishes to a relatively stable plateau. This also suggests that leaching should be higher from columns where the ligand is oriented to provide maximum accessibility for antibody-binding. This is born out by the leakage differential between Pharmacia's wild-type protein A Fast Flow Sepharose and the new recombinant rProtein A Fast Flow media.[166]

Table 9.5. Distribution of monoclonal antibodies. Data from reference 167. Additive discrepancies result from incomplete data.

Species	class	subclass	% of total
Mouse	all	all	88.7%
	IgA	—	0.7%
	IgE	—	0.3%
	IgG	all	69.2%
		1	34.3%
		2a	16.2%
		2b	8.0%
		3	3.0%
	IgM	—	15.7%
Rat	all	all	8.3%
Human	all	all	2.8%

Other indications that proteolysis is the primary leakage pathway include the fact that leaching is often highly elevated in the first run after storage of used media.[152] This makes it imperative to run stored columns though an elution cycle before restoring them to active status.[17,29] Elevated leakage is likewise seen when feedstreams carry high protease loads, such as when there has been a large amount of cell lysis. This can be an important process variable in cases where product performance is affected by leached protein A. Observations of higher leakage with mouse antibodies are also consistent with the proteolysis hypothesis. The higher pH required for binding favors protein A digestion by serine proteases.[12,13,134]

leaching by matrix degradation

A small fraction of the protein A leached from agarose media remains coupled to agarose fragments, indicating breakdown of the glycosidic linkages in the gel media.[17] However, matrix degradation does not appear to be a major leaching pathway for agarose or other polymer-based media. It is a serious limitation for silica or controlled-pore glass media, both of which dissolve at alkaline pH.[35] This is a particular concern for antibodies that require elevated pH to achieve adequate binding. Since the majority of monoclonals are mouse antibodies, and most of those are IgG_1, this is a significant issue (Table 9.5).[167]

leaching by linkage degradation

Degradation of the immobilization linkage has proved not to be a major concern. Even though isourea linkages are widely viewed as a labile—especially at alkaline pH—cyanogen bromide activated media exhibit among the lowest protein A leakage values of all linkage chemistries.[152,153] The unexpected stability of these gels has been attributed to multipoint attachment of the protein A. The implication is that even if some linkages fail, the molecule will remain bound by others. More likely, lateral attachment limits steric accessibility of the interdomain cleavage sites while multipoint linkage limits release of proteolysed fragments. This suggests that multipoint-bound protein A media should maintain binding capacity over their lifetimes better than media with the ligand oriented for higher antibody accessibility. Comparative data are not yet available to test this hypothesis.

the fate of leached protein A

Upon administration in vivo, soluble protein A:IgG complexes dissociate, and the protein A forms new complexes with the IgG of the recipient.[168] Complexes transform similarly in vitro. This raises 2 sets of concerns. The first is the destabilizing effect of transferred protein A on recipient antibody. This may alter patient response or interfere with interpretation of diagnostic results.

immunomodulation by protein A

The second set of concerns derives from protein A being a potent immunomodulator in its own right. In many cases, fragments mediate the same interactions as the intact molecule.[169] Even single binding domains are able to crosslink 2 separate antibodies, apparently through combination of the Fc and Fab binding mechanisms. Table 9.6 lists some of the immunological phenomena known to be affected by protein A.[103,168-251] Expectedly, it has proven toxic in a number of clinical trials.[252-255]

Protein A contamination affects different product applications to different degrees. Many in vitro products are apparently inert, especially low-sensitivity diagnostics. In vitro assays linked in any way to the complement cascade are at high risk, as are assays performed on live mammalian immune cells, platelets, or erythro-

Table 9.6. A partial listing of secondary immunological phenomena mediated by protein A. Many of the findings reported in these publications were obtained by direct addition of protein A to experimental samples and may not be representative of results obtained with protein A leached during IgG purification. Even for those where the results are mediated by leached protein A, critical response thresholds are often not indicated. Nevertheless, these reports clearly identify high-risk applications and provide useful focal points for evaluating leachate effects.

Function	references
Complement activation	115–178
Accelerated IgG catabolism	168
Agglutination of granulocytes	169
Leukocyte chemotaxis	179–184
Induction of hypersensivity	185–191
Histamine release from leukocytes	192
Anaphylaxis	188
Migration of IgG receptors on cell surfaces	193,194
Proliferation of peripheral blood lymphocytes	195–204
Proliferation of T lymphocytes	195–201, 205–208
Proliferation of B lymphocytes	195–201, 205–217
Altered ion transport across lymphocyte membranes	209,210
Stimulation of DNA synthesis in activated B lymphocytes	209
Lymphokine secretion	194, 218,219
Induction of immunoglobulin secretion by B lymphocytes	195,198, 201, 206 215, 217,220,221
Potentiation of natural killer activity of human lymphocytes	222
Inhibition or enhancement of antibody-dependent cell-mediated cytotoxicity	223–227
Modulation of phagocyte function	228–234
Macrophage stimulation	223-226, 235, 236
Induction of rheumatoid factor	219, 237–240
Potentiation of immune response	241–243
Inhibition of binding to Fc receptors	243–247
Inhibition of aggregate binding by lymphocytes	248
Inhibition of lymphocyte colony formation	249
platelet injury	250,251

cytes. Protein A can trigger secondary phenomena completely independent of the antibody. Not only may test results be quantitatively unreliable, they may suggest a diagnosis irrelevant to the patient's actual condition. In vivo applications are also at elevated risk, especially high dosage products or products intended for repetitive application to treat chronic disorders.

Some users speculate that protein A leachate levels are beneath concern. To put the issue in perspective, if protein A was an optional cell culture additive, validation and marketing groups would probably adopt a zero tolerance posture and blacklist it entirely—for in vivo and in vitro products alike. This is not to say that it should be abandoned as a purification tool. Its use in purification of approved products for in vivo application attests that it can be used safely, and its application to numerous in vitro products indicates that its compromises to product performance are generally tolerable. The impact of leaching on product safety and efficacy must nevertheless be qualified rigorously.

removal of leached protein A

It's generally easier to remove leached protein A than to measure it accurately. Removal also relieves the development burden of determining critical leachate thresholds for potentially affected product performance parameters. It similarly relieves the validation burden of documenting that process controls are adequate to maintain leaching within specified ranges. Protein A removal is required for in vivo products in any case.

leachate removal by cation exchange

Cation exchange can achieve quantitative protein A removal from most antibodies, but it usually requires method development.[256] Selectivity variation among cation exchangers, differences in the leachate dissociation requirements for various antibodies, and differences in cation exchange retention properties of individual monoclonals all preclude successful application of generic protocols.

Cation exchange removal of leachate is based on the charge differential between antibodies and protein A. Protein A is aspartate and glutamate rich, with an acidic pI of about 5.1.[15,16,257,258] It binds poorly to cation exchangers. IgGs tend to be alkaline and most

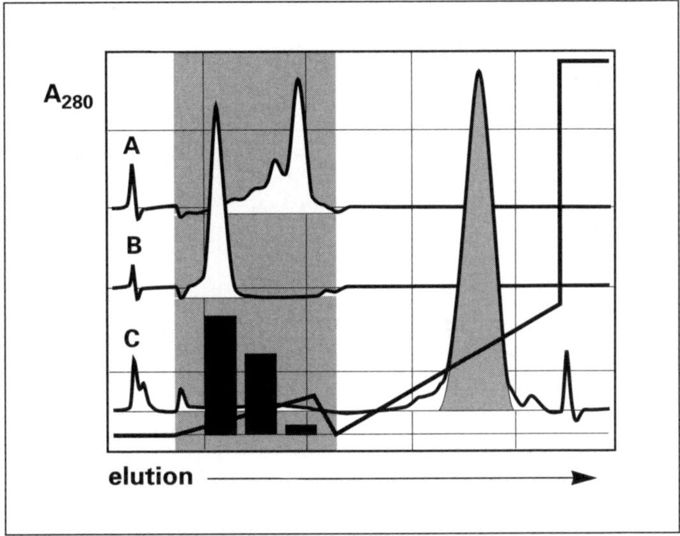

Figure 9.10. Cation exchange removal of leached protein A. A: Commercially obtained purified protein A. B: Recombinant protein A, domain B. C: protein A-purified Mouse IgG_1. The gray zone indicates the protein A-dissociating-eluting wash. The shaded peak indicates the IgG. The dissociating wash is normally applied in a step, but a gradient is used here to illustrate the size-discriminating ability of cation exchange with protein A. The black bars indicate elution of leached protein A. As shown, the majority of the leached protein A is heavily proteolysed. This is typical.

bind strongly. There are 2 approaches for exploiting this differential. The first is to apply the contaminated antibody in a complex-dissociating formulation, under conditions where the protein A is unretained by the cation exchanger. The weakness of this approach is that it may require extended free-solution exposure of the antibody to denaturing conditions.

An equally effective but more conservative approach is to apply the contaminated antibody under "standard" IgG-cation exchange conditions, then apply a protein A-dissociating-eluting wash. Prebinding the antibody has a stabilizing effect. Despite being denatured rapidly at pH 3.0 in free solution, antibodies withstand longer exposure when immobilized.[259] This does not automatically recommend dissociating with extreme pH as a matter of routine. It's important to apply the most moderate conditions possible.

Addition of ethylene glycol and urea disrupt the hydrophobic and hydrogen bonding components of the IgG:protein A interaction, substantially moderating the dissociation pH without elevating risk of denaturation, and without eluting the antibody. Elute a control sample of purified antibody from a protein A column with a pH gradient in 50% ethylene glycol and 2.0M urea. This reveals the pH required for decomplexation. Apply

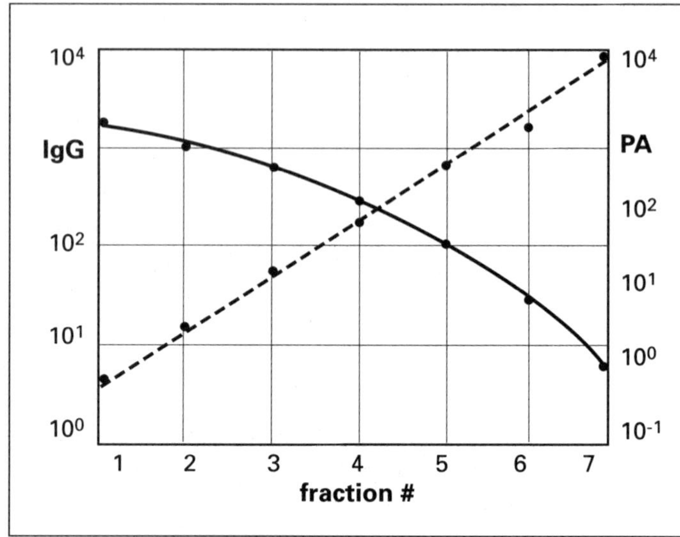

Figure 9.11. Anion exchange reduction of leached protein A. Protein A-contaminated IgG was loaded onto an anion exchanger and eluted with a linear salt gradient. Fractions were collected across the IgG peak and evaluated for protein A. The solid trace indicates IgG in mg. The dashed trace indicates ng protein A per mg of IgG. Data plotted from reference 150.

a sample of purified protein A to the cation exchanger under the IgG:protein A dissociation conditions. If it binds, elute with a linear salt gradient. The salt concentration at the tailing boundary of the protein A peak—in combination with the pH, ethylene glycol, and urea—defines the minimum conditions for leachate dissociation-elution. Apply a sample of purified antibody to the column under these conditions to ensure that it binds. Figure 9.10 illustrates results from a monoclonal applied in acetate buffer at pH 4.5, treated with a protein A-dissociating elution wash, restored to pH 4.5 acetate, then eluted free of protein A in a linear salt gradient. Antibody recovery was >90%.

leachate removal by anion exchange

Anion exchange supports good but inferior leachate removal. The interaction of protein A and IgG is stabilized under the alkaline conditions required for anion exchange of most IgGs. In contrast to antibodies, protein A is very strongly retained; roughly equivalent to albumin. Complexes elute according to their proportional compostion of IgG and protein A, with the most highly protein A-enriched fractions eluting in the tail of the IgG peak (Figure 9.11).[149] This routinely supports >80% protein A removal, seldom with less than 80% recovery of antibody. Quantitative leachate removal by anion exchange is impossible under any circumstances.

leachate removal by HIC

Leachate removal by HIC exploits the differential hydrophobicity of protein A and antibodies. Antibodies are among the more hydrophobic proteins, but protein A is moreso. Complexes exhibit intermediate retention depending on their proportional composition of IgG and protein A. The tailing fractions of the IgG peak are protein A-enriched. Quantitative leachate removal is impossible because of the complex-stabilizing effect of the high salt concentrations. HIC is less effective than anion exchange due to the hydrophobic differential between complexants being less than the charge differential. 50% reduction of protein A may require sacrificing 50% of the antibody.[260]

leachate removal by SEC

SEC under dissociating conditions can also be used to remove protein A, but the lengthy free-solution exposure of the product to the harsh conditions unnecessarily risks denaturation.[69,149,184,261] SEC under nondissociating conditions removes complexes containing at least 2 antibodies but cannot remove monovalent complexes with protein A fragments.[69,261] As with HIC and anion exchange, product is removed along with the protein A.

measuring leached protein A

Measurement of leached protein A is complicated. It is complexed with the antibody, which restricts steric access of whatever ligand is being used in the assay. This requires the use of a standard curve created by combining free protein A with nonaffinity-purified antibody. Developing this curve is itself something of an art since steric accessibility of the protein A depends on the stoichiometry of complexation, which in turn varies with the composition, concentrations, and ratios of the complexants. The situation is complicated further by the fact that there is no way to determine how the complex-distribution in the curve relates to the distribution in your protein A-purified purified sample. This limitation applies equally to prepared standards included with commercial measurement kits.

The impasse can be largely resolved by capturing the protein A-complexed antibody on a cation exchanger, selectively dissociating and eluting the protein A, then measuring it against a pure protein A standard curve. Final determination may still be influenced by the degradation state of the protein A, so run curves for both intact and single-domain fragments.[262] This will limit accuracy to a range of values and there will be a lot-to-lot error factor in your measurements due to variations in protein A composition, but the data will be adequate for most purposes.

measuring effects of leached protein A

Once the level of protein A contamination has been characterized, its effects on the product application can be measured. This is easily done for in vitro diagnostic products by comparing performance of the protein A purified antibody with nonaffinity purified material. Using the nonaffinity purified material, prepare a protein A-contaminated sample containing at least 10 times the measured leachate value in your protein A-purified antibody. This will aid positive identification of low level responses. If you discover a positive response with either your protein A-purified antibody or the 10x contaminated control, prepare samples covering a range of contamination levels so that you can identify the response threshold.

choice of test subjects

Test subjects should include healthy normal individuals, patients representing varying levels of the analyte of interest, and patients with conditions likely to interfere with the assay. The results from this study can be used as the basis to set specifications for allowable protein A content. General guidelines do not exist and are unlikely to evolve. The effectivity ranges for various product applications are so diverse that general guidelines would be meaningless.

testing in vivo products

Testing in vivo or ex-vivo products is more of a challenge. The logistics, time, and expense of conducting paired clinical studies with protein A and nonaffinity-purified antibody are generally regarded as prohibitive. On the other hand, if you're going to invest millions of dollars over a period of years, you need to have assurance that the product won't ultimately be scuttled by having overlooked a critical limitation in your purification process. This is important even if you can prove absolute removal of leached protein A. Protein A-mediated denaturation may still be a problem.

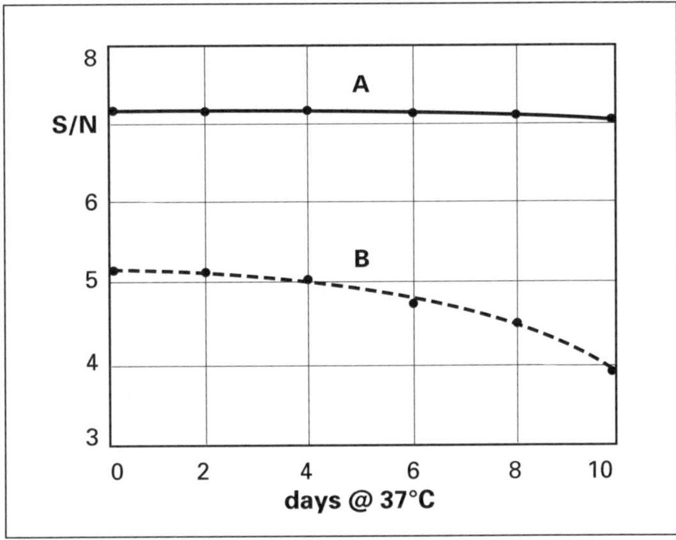

Figure 9.12. Paired stability comparison of a diagnostic reagent made with nonaffinity-purified (A) and protein A-purified mouse IgG$_1$ (B). All values expressed as $\log_{10}$ of specific signal (S) divided by $\log_{10}$ of nonspecific signal (N). S/N for the protein A purified antibody was ~100 times lower than the nonaffinity purified. Its degradation rate was almost 10 times higher. Whether such results are significant for a particular product depends on its application requirements. Some products fulfill their performance needs adequately despite this type of impariment.

Flow cytometry can provide a preliminary determination of product integrity. This technique allows you to quantitatively measure the influence of protein A or its secondary effects on specific immunological mechanisms. It also allows you to test samples representing your intended patient pool. This is important because many recipients of therapy are immunologically challenged and may respond differently than healthy individuals. Use a nonaffinity purified control and be careful to qualify cytometry reagents as nonaffinity-purified. Remarkably, some manufacturers use protein A.

An alternative tool for characterizing the effects of protein A is microphysiometry.[263-266] This technique has not yet been applied to measurement of protein A-mediated cytotoxicity, but it offers the capability of providing continuous measurement over time.

stability testing

Whether your product can tolerate leached protein A or whether it has to be removed, it's good insurance to run accelerated thermal degradation studies to make sure that subtle protein A-mediated alterations don't escape detection (Figure 9.12). Look for loss of antibody titer, elevated nonspecific interactions, and reduced stability (Figures 9.7 & 9.12). Testing a representative variety of patient samples will help avoid surprises during clinical trials or in the marketplace.

immobilized impurities

Leaching of ligand contaminants derived from the protein A source material are another concern. Especially when the protein A is isolated from Staphylococcal cell lines, copurification of pathogenic agents can occur. The T-cell activation and lymphokine secretion attributed to protein A are often caused by Staphylococcal enterotoxin A.[195,196,210,211] This particular contaminant problem doesn't afflict recombinants, but high ligand purity is still important.

contaminants from purification of the protein A

Contaminants from the protein A purification process itself are another issues, regardless of whether the protein A was produced in Staphylococci or non-pathogenic sources. Chief among these are viruses that may leach from the human IgG columns often used to purify protein A. Even if the IgG is tested virus-free, the posture of regulatory agencies is that testing can't demonstrate absence of a virus you didn't specifically test for. Some suppliers have already responded to this risk by ceasing to purify their protein A with immobilized human IgG.[166]

variations among immobilized protein A media

All protein A media exhibit the same qualitative response to the same chemical stimuli; for example, stronger antibody binding at elevated pH and salt concentration. However they do not respond to the same degree. Some media present the ligand in a way that is believed to allow protein A binding to the Fc sites on both sides of the IgG.[35] The net association/dissociation constants change significantly. Weakly-binding IgGs are able to adsorb under more moderate conditions and elution requires more extreme conditions. This can be useful or counterproductive, depending on your product and its application requirements.

An important qualification to be aware of with media that support "dual-site" binding, is that not all the antibodies may be bound by both sites. Under buffer conditions that support single-site binding, if you load the column beyond its dual-site binding capacity, you will create a subpopulation of single-site bound antibody. This has important consequences. If you apply a homogenous monoclonal population, it will elute in separate peaks under 2 different sets of

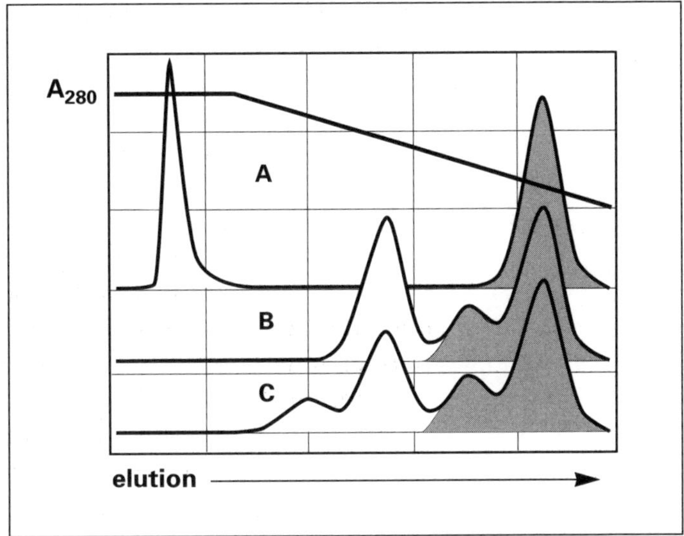

Figure 9.13. Single and dual-site binding as a function of buffer conditions. Shaded areas indicate Human IgG. Unshaded areas indicate Mouse IgG. Profile A illustrates conditions that support only dual-site binding of human IgG. Profile B illustrates conditions that support single and dual-site binding of Human IgG, but only dual site binding of Mouse IgG. Profile C illustrates conditions that support single and dual-site binding of both. See text for discussion.

conditions. If you apply a raw mixture of a specific monoclonal with contaminating polyclonals, fragments, and aggregates, overlap of elution conditions among the various single- and dual-site bound populations will compromise purification performance.

The effects of mixed-mode elution are diagrammed in Figure 9.13 where a hypothetical mixture of Mouse and Human IgG_1 is loaded to breakthrough capacity under various conditions, then eluted in a linear pH gradient. In addition to depressing product purity, mixed-mode elution can affect recovery. If step elution conditions are set to remove weakly-bound contaminating antibodies in advance of the main body of product, single-site bound product can be lost with the contaminants.The insidious aspect of this phenomenon is that it's easy to overlook during early method development. It doesn't become apparent until you load the gel beyond dual-site dynamic capacity.

There are 2 ways to exploit a gel's stronger binding capabilities without sacrificing resolution. The simplest is to develop buffer conditions that support only dual-site binding. You can estimate the upper limit of the range by evaluating product-binding requirements on a gel limited to single-site binding. These gels can usually be identified by Human IgG capacities ≤20mg/mL of gel. Use the strongest promoting conditions that don't support retention on the single-site gel.

The alternative approach can be used regardless of binding conditions. You need to define the product load limits above which the product begins to exhibit single-site binding (double-peak elution). Load the column with an increment of product and elute in a linear pH gradient. Double the load and repeat until you reach a point where the product begins to elute in 2 peaks. This indicates that you have exceeded its dual-site dynamic capacity. Fine-tune the load until you restore single peak elution.

Although dual-site binding is presently known to occur with products from only one supplier, it likely affects all high ligand-density media where the protein A is immobilized to project out from the matrix surface. Dynamic capacity measurements on protein A should therefore always be accompanied by linear gradient elution to reveal potential compromises to purification performance. The exception is if your sole interest is capacity and you have no requirement to fractionate the product from contaminating antibodies, fragments, or aggregates. In that case, set loading conditions to support single-site binding and measure dynamic capacity according to the frontal analysis method described in chapter 4.

column equilibration requirements

Protein A columns require the same extensive equilibration as ion exchangers, for the same basic reason: pH titration of the ligand—in this case the critical histidyl residues. Inadequate titration amplifies process variation and may falsely indicate that a particular antibody doesn't bind. Figure 9.14 illustrates the pH titration curve of a protein A column previously eluted at low pH. This type of characterization is an essential part of method development for all protein A applications.

cross-contamination

The difficulty of removing strongly bound antibodies imposes an important limitation on protein A media used for commercial purifications. It restricts a particular column to a single product. The risk of cross-contamination is too high to support multi-product application. Protein A's resistance to physicochemical stress

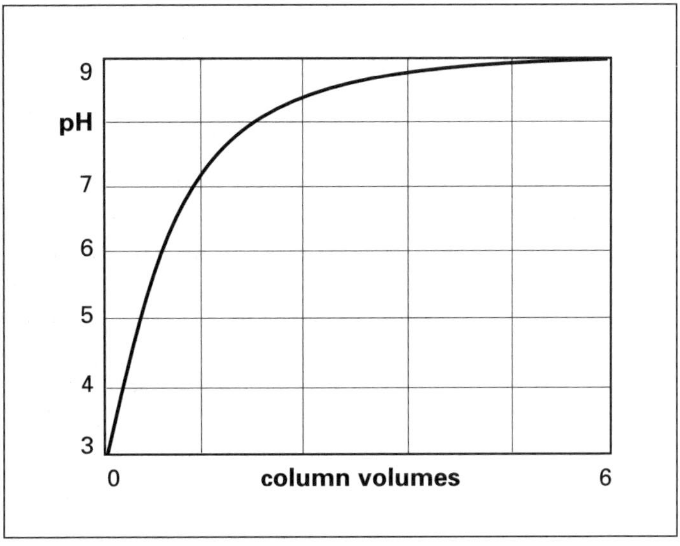

Figure 9.14. pH titration of a protein A column after elution at pH 3.0. The volume of buffer required for titration is variable depending, on the column, choice of buffering compounds , the pH of the column before equilibration is commenced, and other factors. The only way to determine column equilibration requirements is by direct experimentation.

makes it possible to develop and validate cleaning procedures that prevent cross-contamination. However, convincing regulators tends to require a level of overkill that significantly reduces media life.

antibody solubility limitations

Antibodies with weak protein A affinities are likely to be partially or wholly insoluble in the high salt concentrations required to achieve worthwhile dynamic capacities. This precludes advance off-line batch equilibration. As with HIC, equilibrating and loading the sample by on-line dilution solves the problem (chapter 6).

temperature effects

Temperature is a significant variable in protein A performance.[59] Shifting a room-temperature process to a coldroom radically alters antibody binding and elution behavior. Relatively modest temperature variations between the development lab and the manufacturing floor are a frequent source of scale-up aberrations. Similarly uncontrolled variations in manufacturing amplify process variation. Make sure that temperature in the manufacturing area is well controlled and that protein A processes are developed at the same temperature as the facility for which they are destined.

buffer incompatibilities

An occasional affliction with IgG_3s and other strongly basic antibodies is that some are sufficiently charged to form stable crosslinks with polyvalent anions. These include phosphate, citrate, sulfate, and

Table 9.7. Screening buffers and conditions. Use buffer A1 for binding Human and Guinea Pig IgG. Use buffer A2 for all others. The recommended column equilibration interval is excessive under most conditions, but should be used as a default until specific equilibration requirements are established for your particular system.

Buffers

A1: 0.02 M sodium phosphate, 0.02M sodium citrate, pH 7.5

A2: 0.02M boric acid, 0.02M sodium phosphate, 0.02M sodium citrate, 1.0M sodium sulfate, pH 9.0

B: 0.02M sodium citrate, 0.1M sodium chloride, pH 2.5

Conditions

Equilibrate column: with 10CV buffer A
Inject: 5%CV unequilibrated sample
Wash: 2CV buffer A
Elute: in a 10CV linear gradient to buffer B.
Strip: with a 5CV buffer B.

borate; some of the most commonly used protein A buffer components. The precipitates formed by this interaction depress purification performance, especially recovery. If you encounter one of these antibodies, use organic buffers and monovalent salts.

cost issues

The price of protein A media is many times the cost of nonaffinity supports. This ripples through the purification process and elevates cost per unit of product. Cost per gram of a mouse IgG_1, based on a 10-cycle comparison, was >4 times higher for 1-step protein A purification than for a 2-step nonaffinity process. [112] Differentials vary according to raw materials and procedures, but the relative cost burden of protein A is substantial in all cases.

column maintenance

The high density of controlled pore glass protein A media is sometimes suggested to be disadvantageous, on the pretense that they are difficult to unpack. Tools that may fracture the particles should be avoided but used media can be removed easily by flowing buffer from the bottom of the column. This fluidizes the bed so that the media can be pumped or decanted directly into another container. This works even on columns that have been packed for years.[267]

Method development

Protein A method development is complicated by its variable IgG affinity; the need to remove nonspecific

Table 9.8. pKs of phosphate and citrate buffers. Data from reference 167.

Buffer		
phosphoric acid	pK_1	2.15
citric acid	pK_1	3.13
citric acid	pK_2	4.76
citric acid	pK_3	6.40
phosphoric acid	pK_2	7.20
phosphoric acid	pK_3	12.33

antibodies, fragments, aggregates, and leached protein A; and the need to minimize denaturation. The IEC method development model discussed in chapter 4 provides a systematic framework for addressing these issues. Only points of difference are discussed here.

screening conditions

Although step gradients are preferred in most final processes, it's important to conduct initial selectivity screening with linear gradients (Table 9.7). Foreknowledge of a monoclonal's subclass may suggest a particular range of conditions, but variation from one monoclonal to another is sufficient that there is a risk of either splitting the peak or missing it altogether.[268] Rather than hoping for a lucky hit, it's more efficient to characterize elution characteristics, then secondarily customize conditions to accommodate them. It may also reveal worthwhile fractionation opportunities.

gradient formation

The key point in developing reproducible linear pH gradients is even buffering capacity over the selected pH range. Phosphate/citrate systems are well suited to this application. Both are anionic and the combination of their pKs provides a ladder over the range from pH 7.2 to 2.2 (Table 9.8).[47] This supports stable reproducible gradients from ~pH 8.0 down to 2.0. If a higher pH range is required, add boric acid to the binding buffer. Borate is also anionic with a pK of 9.2. Gradient generation consists of simple proportional mixing of the high pH buffer with the low pH buffer.

sample re-equilibration

If your antibody elutes below pH 4.5, it's usually worthwhile to evaluate the pH-moderating effects of elevated salt concentration, urea, and ethylene glycol. If your final product must be free of leached protein

A, data derived from this evaluation are also useful for defining dissociating wash conditions for cation exchange chromatography.

gradient format

Linear gradients support the best separation of the product from nonspecific antibodies, fragments and aggregates; also the best reproducibility versus uncontrolled process variations. These attributes can be especially important if you are relying on protein A as a 1-step purification. Step gradients yield higher product concentration, especially if the direction of flow is reversed during elution. Flow reversal can also improve recovery by up to 5%.

sample re-equilibration

Develop sample re-equilibration conditions by titrating the antibody fraction with 1.0M potassium phosphate or Tris, pH 7.5 until the pH reaches 6.0. Calculate the relative volume of buffer added and specify that this proportion be prespiked into fraction collection vessels prior to elution.

leachate removal

If leached protein A is an issue, evaluate it after you have established dynamic capacity and have representative samples from several lots to compare. You can save work by removing the protein A instead. Assuming you are able to document quantitative removal, this will suspend validation requirements for characterization of leaching under various process conditions.

scale-up issues

One of the most common sources of scale-up problems with protein A is failing to note and accommodate the solubility limitations of antibodies that require high salt concentrations for binding. Characterize product solubility throughly under loading conditions. If it fails to remain fully soluble for the longest possible duration from equilibration to completion of sample load, then load by on-line dilution.

Another frequent problem is inadequate equilibration of column pH before loading. Once the elution and binding buffers are established, equilibrate the column with 10 column volumes of elution buffer then measure the volume of binding buffer required to bring the pH to binding specifications.

Inadequate attention to elution and re-equilibration conditions causes unnecessary variations in product performance. Too many monoclonals are denatured by exposure to unnecessarily extreme conditions and failure to re-equilibrate the product upon elution. Symptoms include depressed and erratic product performance, and poor stability.

Uncharacterized variation of leachate contamination due to variability of feedstream composition also affects product performance—in a far wider diversity of products than most users initially suspect. Poor column hygiene, especially when paired with omission of a pre-run elution step, compounds the problem.

In perspective

Protein A currently ranks with IEC as one of the most popular methods for antibody purification. This is reflected by the wide range of product application formats available to process designers. One of the most interesting is represented by media that support direct capture from crude process streams without previous centrifugation, prefiltration, or concentration.[269-271] These particular products are designed mainly for large scale applications, but a variety of other high-throughput media offer good process economy to people working at more modest process scales.

purification of in vitro products

Protein A is at its best in the field of in vitro diagnostics, where mouse antibodies dominate. Elution conditions can be developed that minimize antibody denaturation, and 1-step purity is adequate for many applications. Combination with cation exchange removes most nonspecific antibody and can be optimized to remove leached protein A at the same time. Combination with anion exchange provides similar capability for removal of nonspecific antibodies, and significant—though inferior—leachate removal.

purification of in vivo products

Protein A is also used successfully for commercial purification of in vivo products. Its ability to provide excellent clearance of DNA, virus, and endotoxin figures strongly in its favor. Supporting methods are still required to achieve regulatory standards for these contaminants, and to remove nonspecific antibodies and leached protein A. Consequently, protein A-based procedures typically include as many steps as nonaffinity schemes. Such procedures have proved to be competitive with exclusively nonaffinity systems, but any advantages they may provide in terms of higher initial purity are offset by elevated material and validation costs.

Many of the current generation of protein A-purified in vivo products benefit from being mouse-derived; able to support dissociation of protein A under relatively mild conditions. A growing majority of the newer products are human IgGs or chimeras with human Fc regions. This translates into harsher elution conditions, higher probability of denaturation, and greater difficulty removing leachate. Whether you approach it from a regulatory, product performance, or economic perspective, this makes protein A more of a liability than an asset. Especially for these antibodies, substituting IMAC for protein A preserves the key process advantages while suspending its chief limitations.

Recommended reading

The best point of entry into the protein A literature is Michael Boyle's 2 volume series, Bacterial Immunoglobulin-Binding Proteins.[272] This is an outstanding reference, giving broad coverage to the biology and applications of protein A. Chapters 18 and 19 in volume 2 both specifically address purification. The article by Pepper provides many useful practical tips.[29] The review by Forsgren is a good place to begin developing an appreciation for proteins A's effects on mammalian immunophysiology.[103]

References

1. H. Hjelm et al, 1972, *FEBS Lett.*, **28** 73
2. G. Kronvall et al, 1973, *J. Immunol.*, **111** 141
3. J. Björck et al, 1972, *Eur. J. Biochem.*, **29** 579
4. J. Sjödahl, 1977, *Eur. J. Biochem.*, **78**, 471
5. T. Moks et al, 1986, *Eur. J. Biochem.*, **15** 637
6. M. Uhlén et al, 1984, *J. Biol. Chem.*, **259** 1695
7. S. Löfdahl et al, 1983, *Proc. Nat. Acad. Sci. USA*, **80** 697
8. D. Colbert and A. Anilionus, 1983, European patent publication 0107509 A2
9. B. Gus et al, 1985, *Eur. J. Biochem.*, **153** 579
10. J. Diesenhofer et al, 1978, *Hoppe-Seylar's Z. Physiol. Chem.*, **359** 975
11. H. Helm et al, 1978, *Eur. J. Biochem.*, **57** 395
12. J. Sjödahl, 1976, *FEBS Lett.*, **67** 62
13. J. Sjödahl, 1977, *Eur. J. Biochem.*, **73** 343
14. M. Uhlén et al, 1983, *Gene*, **23** 369
15. J. Langone, 1982, *Adv. Immunol.*, **32** 157
16. R. Lindmark et al, 1977, *Eur. J. Biochem.*, **74** 623

17. L. Schwartz, 1990, in Bacterial Immunoglubulin-Binding Proteins (M. Boyle, ed.), Vol. II, p. 309, Academic press, San Diego
18. —1995, STREAMLINE rProtein A, data file 18-1115-67, Pharmacia Biotech AB, Uppsala, Sweden
19. C. Lowe et al, 1995, Purification of Immunoglobulins Using a Novel Synthetic Protein A ligand, poster, The Affinity Technology Conference, Waltham, MA USA, reprint: Prometic Biosciences Inc., Burtonsville, MD
20. Mimetic-A, Prometic Biosciences, Inc., Burtonsville, MD
21. B. Solomon et al, 1992, *J. Chromatogr.*, **597** 257
22. P. Bonnett, 1996, personal communication, Prometic Biosciences Inc., Burtonsville, MD, USA
23. J. Diesenhofer, 1981, *Biochemistry*, **20** 2361
24. M. North, 1989, in Proteolytic Enzymes, A Practical Approach, (R. Beynon and J. Bond, eds.), p.105, IRL Press, Oxford UK
25. J. Pringle, 1975, in Methods in Cell Biology, (D. Prescott, ed.), Vol. 12, p. 149, Academic Press, New York
26. M.-J. Schmerr et al, 1985, *Mol. Immunol.*, **22** 613
27. M.-J. Schmerr et al, 1985, *J. Chromatogr.*, **326** 275
28. H. Juarez-Salinas and G. Ott, 1987, US patent 4,704,366
29. D. Pepper, 1990, in Laboratory Methods in Immunology, (H. Zola, ed.),Vol.II, p. 169, CRC Press, Boca Raton
30. F. Hofmeister, 1888, *Arch. Exp. Pathol. Pharmakol.*, **24** 247
31. P. Gagnon and E. Grund., 1996, *BioPharm*, **9**(5) 54, reprint: <http://www.validated.com/library.html>
32. H. Schultze and J. Heremans, 1966, Molecular Biology of Human Proteins, Vol. I, Elsevier, New York
33. Y. Kato et al, 1984, *J. Chromatogr.*, **298** 407
34. S. Hjerten et al, 1986, *J. Chromatogr.*, **359** 99
35. —1996, Optimization and Scale-up of Therapeutic Antibody Production Using PROSEP-A High Capacity, Bioprocessing Ltd., Consett, UK
36. M. Mackenzie et al, 1978, *J. Immunol.*, **120** 1493
37. P. Gagnon, Multiple Mechanisms for Improving Binding Efficiency with Protein A, poster, BioEast '92, Washington D.C., reprint: <http://www.validated.com/library.html>
38. T. Arakawa, 1985, *Anal. Biochem.*, **144** 267
39. J. Lee and L. Lee, 1979, *Biochemistry*, **18** 5518
40. J. Lee and L. Lee, 1981, *J. Biol. Chem.*, **156** 625
41. T. Arakawa and S. Timasheff, 1983, *Biochemistry*, **23** 5912
42. T. Arakawa and S. Timasheff, 1981, *Biochemistry*, **21** 6545
43. L. Lee and J. Lee, 1987, *Biochemistry*, **26** 7813
44. J. Porath and N. Ui, 1964, *Biochim. Biophys. Acta*, **90** 324
45. E. Kabat and M. Mayer, 1966, Experimental Immunochemistry, 2nd Edition, p. 264, Charles C. Thomas Publisher, Springfield
46. U.-B. Hansson and E. Nilsson, 1973, *J. Imunol. Met.*, **2** 221

47. D. Perrin and B. Dempsey, 1974, Buffers for pH and Metal Ion Control, Chapman and Hall, New York
48. R. Byewater et al, 1983, *J. Immunol. Met.*, **64** 1
49. L. Haff, 1981, *Electrophoresis*, **2** 87
50. M. Yarmush and W. Olson, 1988, *Electrophoresis*, **9** 11
51. E. Kabat et al, 1987, Sequences of Proteins of immunologiocal Interest, US Dept. of Health and Human Services, Public Health Service, National Institute of Health
52. C. Bigelow and M. Channon, 1976, in Handbook of Biochemistry and Molecular Biology, 3rd Edition, Proteins, Vol. I, p. 209, CRC Press, Boca Raton
53. D. Burton, 1985, *Mol. Immunol.*, **22** 161
54. D. Haake et al, 1982, *J. Immunol.*, **129** 190
55. T. Michaelson et al 1977, *J. Immunol.*, **119** 558
56. C. Wolfenstein-Todel et al, 1976, *Biochem. Biophys. Res. Commun.*, **71** 907
57. G. Kronvall and R. Williams, 1969, *J. Immunol.*, **103** 828
58. G. Kronvall et al, 1979, *J. Immunol.*, **104** 140
59. G. Schuler and M. Reinacher, 1991, *J. Chromatogr.*, **587** 61
60. S. Goheen and S. Englehorn, 1984, *J. Chromatogr.*, **317** 55
61. P. Gagnon et al, 1995, *BioPharm*, **8**(4) 36, reprint: <http://www.validated.com/library.html>
62. M. Brunda et al, 1979, *J. Immunol.*, **123** 1457
63. A. Grov, 1975, *Acta pathol. Micobiol. Scand. Sect. C*, **84C** 71
64. A. Grov, 1975, *Acta pathol. Micobiol. Scand. Sect. C*, **83C** 173
65. J. Zikan, 1980, *Folia Biol.*, **25** 254
66. J. Biewenga et al, 1982, *Immun. Commun.*, **11**(3) 189
67. H. Juarez-Salinas, 1987, in Commercial Production of Monoclonal Antibodies (S. Seaver, ed.), p.217, Marcel Dekker, New York
68. M. Mariani et al, 1989, *Immunol. Today*, **10** 115
69. M. Mariani et al, 1989, *BioChromatography*, **4** 149
70. M. Harboe and I. Folling, 1974, *Scand. J. Immunol.*, **3** 471
71. M. Mackenzie et al, 1978, *Scand. J. Immunol.*, **71** 367
72. I. Lind et al, 1975, *Scand. J. Immunol.*, **4** 843
73. A. Milon et al, 1978, *Dev. Comp. Immunol.*, **140** 1223
74. S. Biguzzi et al, 1982, *Scand. J. Immunol.*, **15** 605
75. C. Endresen, 1979, *Acta Pathol. Microbiol. Scand. Sect. C*, **87C** 185
76. R. Lindmark et al, 1983, *J. Immunol. Met.*, **62** 1
77. G. van Kamp, 1979, *J. Immunol. Met.*, **27** 301
78. E. Saltvedt and M. Harboe, 1976, *Scand. J. Immunol.*, **5** 1103
79. E. Howell-Saxton and F. Wettstein, 1978, *J. Immunol.*, **121** 1334
80. M. Ingänäs, et al, 1980, *Scand. J. Immunol.*, **12** 23
81. M. Ingänäs, 1981, *Scand. J. Immunol.*, **13** 343
82. M. Boyle, 1990, in Bacterial Immunoglobulin-Binding Proteins, (M. Boyle, ed.), Vol. I, p. 17, Academic Press, San Diego

83. M. Ingänäs and S. Johanson, 1981, *Int. Archs. Allerg. Appl. Immun.*, **65** 91
84. R. Duhamel et al, 1979, *J. Immunol. Met.*, **31** 211
85. S. Ito et al, 1980, *Proc. Jpn. Acad.*, **56** 226
86. B. Recht et al, 1981, *J. Immunol. Met.*, **127** 917
87. E. van Loghem, 1982, *Scand. J. Immunol.*, **15** 275
88. P. Ey et al, 1978, *Immunochemistry*, **15** 429
89. I. Sepälä et al, 1981, *Scand. J. Immunol.*, **14** 337
90. M. Chapman et al, 1979, *Scand. J. Immunol.*, **9** 359
91. L. Martin, 1982, *J. Immunol. Met.*, **52** 202
92. J. Goding, 1978, *J. Immunol. Met.*, **20** 241
93. G. Medgyesi et al, 1978, *Immunochemistry*, **15** 125
94. J. Goudswaard et al, 1978, *Scand. J. Immunol.*, **8** 21
95. M. Ricardo et al, 1981, *J. Immunol.*, **127** 946
96. T. Miller and H. Stone, 1978, *J. Immunol. Met.*, **24** 111
97. H. Hjelm, 1975, *Scand. J. Immunol.*, **4** 633
98. K. Nilsson et al, 1982, *Mol. Immunol.*, **19** 119
99. J. Rosseaux et al, 1981, *Mol. Immmunol.*, **18** 639
100. M. Watanabe et al, 1981, *Jpn. J. Exp. Med.*, **51** 65
101. S. Fulton et al, 1991, *BioTechniques*, **11**(2) 226
102. G. Lindahl and G. Kronvall, 1988, *J. Immunol.*, **140** 1223
103. A. Forsgren et al, 1983, in Staphylococci and Staphylococcus Infections, (C. Easmon and C. Adlam, eds.) Vol. II, p. 428, Academic Press, London
104. P. Underwood et al, 1983, *J. Immunol., Met.*, **60** 33
105. D. Nau, 1988, in Nucleic Acid and Monoclonal Antibody Probes, (B. Swaminathan and G. Prakash, eds.) Marcel Dekker, New York
106. D. Nau, 1987, in Commercial Production of Monoclonal Antibodies, (S. Seaver, ed.) Marcel Dekker, New York
107. D. Nau, 1986, *BioChromatography*, **1**(2) 82
108. M. Gemski et al, 1985, *BioTechniques*, **3**(5) 378
109. T. Brooks and A. Stevens, 1985, *Am. Lab.*, **Oct** 54
110. C. Prior in Animal Cell Bioreactors, (C. Ho and D. Ng, eds.) p. 445, Butterworth, New York
111. P. Underwood et al, 1984, *J. Immunol. Met.*, **60** 33
112. R. Leatherbarrow and R. Dwek, 1983, *FEBS Lett.*, **164** 227
113. M. Nose and H. Wigzell, 1983, *Proc. Nat. Acad. Sci. USA*, **80** 6632
114. L. Björck and B. Åkerström, 1990, in Bacterial Immunoglobulin-binding Proteins, (M. Boyle, ed.) Vol. I., p.113, Academic press, San Diego
115. S. Ohlsen et al 1988, *J. Immunol. Met.*, **114** 231
116. B. Gus et al, 1990, in Bacterial Immunoglobulin-Binding Proteins, (M. Boyle, ed.), Vol. I, p. 21, Academic Press, San Diego
117. —1990, Trio Product Literature, Sepracor Inc. (now BioSepra SA, France)
118. J. Levy et al, 1984, *Clin. Exp. Immunol.*, **56** 114

119. S.-M. Lee, 1986, *J. Biotechnol.*, **4** 189
120. P. Gagnon, 1996, Special Weapons and Tactics for Removing Product-Bound DNA, slide presentation, BioEast '96, Washington, D.C.
121. J. Grun et al, 1992, *BioPharm*, **5**(9) 22
122. M. Rhodes, 1990, *Gen. Eng. News*, **March** 7
123. S.-M. Lee, 1987, in Commercial Production of Monoclonal Antibodies, S. Seaver, ed.), p. 199, Marcel Dekker, New York
124. S. Scott and H. Juarez-Salinas, 1990, in Bacterial Immunoglobulin-Binding Proteins, (M. Boyle, ed.), Vol. II, p.341, Academic Press, San Diego
125. —1994, Protein A Sepharose QuickFacts, publication 18-1060-30, Pharmacai Biotech, Uppsala
126. M. Asplund et al, 1996, Cleaning in Place of STREAMLINE rProtein A, poster, Cell Culture Engineering V., San Diego
127. A. Kenney and H. Chase, 1987, *J. Chem. Technol. Biotechnol.*, **39** 173
128. R. Broeze et al, 1996, Development of Production and Purification Processes for a Monoclonal-based Imaging Product, slide presentation, Waterside Monoclonal Conference, Norfolk, VA USA
129. S. Duffy et al, 1989, *BioPharm*, **2** 34
130. R. Dufff, 1985, *Trends. Biotechnol.*, **3** 167
131. R. Alan and H. Isliker, 1974, *Immunochemistry*, **11** 175
132. R. Alan and H. Isliker, 1974, *Immunochemistry*, **11** 243
133. J. Winkehake, 1978, *Immunochemistry*, **15** 695
134. R. Parehk, 1992, *Biotech. Lab.*, **11** 61
135. Y. Kagawa, 1988, *J. Biol. Chem.*, **263** 508
136. M. Malaise, 1987, *Clin. Immunol. Pathol.*, **45** 1
137. B. Maiorella et al, 1993, *Bio/Technology*, **11**(3) 387
138. T. Monica et al, 1993, *Bio/Technology*, **11** (4) 512
139. C. Goochee et al, 1990, *Bio/Techniques*, **8**(5) 421
140. C. Goochee et al, 1992, in Frontiers in Bioprocesing II, (P. Todd et al, eds.) p.129, American Chemical Society, Washington D.C.
141. R. Beynon and J. Bond (eds.), 1989, Proteolytic Enzymes, A Practical Approach, IRL Press, Oxford, UK
142. B. Robert and R. Bockman, 1967, *Biochem. J.*, **102** 554
143. C. Tanford,1968, *Adv. Protein Chem.*, **23** 121
144. S. Gerlsma and E. Stuur, 1974, *Int. J. Peptide Protein Res.*, **6** 65
145. S. Budvari (ed.), 1989, The Merck Index, 11th ed., Merck and Co. Inc., Rahway, NJ
146. C. Middaugh and G. Litman,1977, *J. Biol. Chem.*, **252** 8002
147. W. Jencks, 1969, Catalysis in Chemistry and Enzymology, p.323, McGraw-Hill, New York

148. S. Timasheff and G. Fasman, 1969, Structure and Stability in Biological Macromolecules, Marcel Dekker, New York
149. G. Sofer and L.-E. Nyström, 1991, Process Chromatography, A Guide to Validation, Academic Press, San Diego
150. J. Bloom et al, *J. Immunol. Met.*, **117** 83
151. R. Francis et al, 1990, in Separations For Biotechnology, (D. Pyle, ed.), Vol. II, p. 491, Elsevier, London
152. M. Godfrey et al, 1992, *J. Immunol. Met.*, **149** 21
153. P. Fugliställer, 1989, *J. Immunol. Met.*, **124** 171
154. H. Griffiths et al, 1993, J. Cellul. Biochem., Protein Purification and Biochemical Engineering, 22nd keystone Symposium, Wiley-Liss, New York
155. —1992, Protein A Contaminant Assay, Threshold Application Notes, publication 0120-326A2, Molecular Devices Inc., Sunnyvale, CA USA
156. G. Salvasan and H. Nagase, 1989, in Proteolytic Enzymes, A Practical Approach, (R. Beynon and J. Bond, eds.) p. 83, IRL Press, Oxford, UK
157. G. Virella and C.-J. Yeh, 1977, *Experentia*, **33** 1231
158. R. Austin and B. Hayward, 1960, *Nature*, **202** 5632
159. C. James et al, 1964, *Nature*, **187** 129
160. J. Sjöquist et al, 1972, *Fedn. Proc.*, **21** 35
161. G. Connick and R. Partner, 1966, *Can. J. Biochem.*, **44** 371
162. G. Virella, 1971, *Experentia*, **27** 94
163. F. Skvaril et al, 1976, *Vox Sang.*, **30** 334
164. G. Davies and J. Lowe, 1963, *Br. J. Pharmacol.*, **21** 491
165. D. Campbell and T. Boenisch, 1964, *Clin. Chem.*, **10** 600
166. —1996, rProtein A Sepharose Fast Flow, Data file 18-1113-94, Pharmacia Biotech Ab, Uppsala, Sweden
167. —1992, American Type Culture Collection Catalog, 7th Ed., American Type Culture Collection, Rockville, MD
168. S. Dima et al, 1983, *Eur. J. Biochem.*, **13** 605
169. J. Barnett-Foster et al, 1982, *Mol. Immunol.*, **19** 247
170. J. Sjöquist and G. Stalenheim, 1969, *J. Immunol.*, **103** 467
171. M. Boyle, 1990, in Bacterial Immunoglobulin-Binding Proteins, (M. Boyle, ed.), Vol. II, p. 369, Academic Press, San Diego
172. G. Kronvall and H. Gewurtz, 1970, *Clin. Exp. Immunol.*, **7** 261
173. G. Stalenheim et al, 1973, *J. Immunochem.*, **10** 501
174. M. Laky et al, 1985, *Mol. Immunol.*, **22** 1297
175. J. Langone et al, 1978, *J. Immunol.*, **121** 327
176. J. Langone et al, 1978, *J. Immunol,*. **121** 333
177. J. Langonc et al, 1984, *J. Immunol.*, **133** 1057
178. J. Balint and F. Jones, 1983, *Immunol. Commun.*, **12** 573
179. J. Lawman et al, 1990, in Bacterial Immunoglobulin-Binding Proteins, (M. Boyle, ed.) Vol.II, p. 405, Academic Press, San Diego

180. J. Lawman et al, 1984, *J. Immunol. Met.*, **69** 197
181. M. Boyle et al, 1984, in Inflammatory Bowel Disease: Current and Future Approaches, (R. McDermot, ed.) p. 371, Excerpta Medica, Amsterdam
182. R. Harvey et al, 1970, *Proc. Soc. Exp. Biol. Med.*, **135** 453
183. A. Sulica, 1979, *Immunology*, **38** 173
184. J. Balint et al, 1984, *Cancer Res.*, **44** 734
185. K. Jensen, 1961, *Acta Path. Microbiol. Scand.*, **53** 191
186. T. Löfkvist, 1966, *Int. Arch. Allergy Appl. Immunol.*, **116** 253
187. G. Gustafson et al, 1967, *J. Immunol.*, **98** 1178
188. G. Gustafson et al, 1968, *J. Immunol.*, **100** 530
189. C. Easmon and A. Glynn, 1975, *Immunology*, **29** 75
190. P. Heczko et al, 1973, *Acta Path. Microbiol. Scand., Sect. B.*, **81** 731
191. J. Bolen and J. Tribble, 1979, *Immunology*, **3** 809
192. A. Petersson and G. Stalenheim, 1975, *Scand. J. Immunol.*, **4** 103
193. V. Ghetie et al, 1974, *Immunology*, **26** 1081
194. L. Räsänen et al, 1978, *Cell. Immunol.*, **37** 221
195. D. Barrett, 1990, in Bacterial Immunoglobulin-Binding Proteins, (M. Boyle, ed.) Vol.I, p. 279, Academic Press, San Diego
196. D. Barrett, 1990, in Bacterial Immunoglobulin-Binding Proteins, (M. Boyle, ed.) Vol.II, p.393, Academic Press, San Diego
197. S. Romagnani et al, 1978, *Immunology*, **35** 471
198. W.-Y. Chen et al, 1982, *Infect. Immunol.*, **36** 59
199. A. Forsgren et al, 1976, *Eur. J. Immunol.*, **6** 207
200. R. Schurmann et al, 1980, *J. Immunol.*, **125** 820
201. O. Ringden and B. Rynell-Dagöö, 1978, *Eur. J. Immunol.*, **8** 47
202. S. Romagnani et al, 1985, *Cell. Immunol.*, **90** 52
203. V. Ghetie et al, 1986, *Mol. Immunol.*, **23** 377
204. R. Carlsson, et al, 1984, *Cell. Immunol.*, **81** 287
205. A. Forsgren et al, 1988, *Cell. Immunol.*, **112** 78
206. O. Saiki and P. Ralph, 1981, *J. Immunol.*, **127** 1044
207. S. Romagnani et al, 1984, *J. Immunol.*, **132** 566
208. T. Kasahara et al, 1980, *Cell. Immunol.*, **49** 142
209. E. Jakobsen et al, 1992, *J. Immunol. Met.*, **152** 49
210. H. Clevers et al, 1986, *Proc. Nat. Acad. Sci. USA*, **83** 4494
211. H. Schrezenmeier and B. Fleischer, 1987, *J. Immunol. Met.*, **105** 133
212. H. Dosch et al, 1980, *J. Immunol.*, **125** 827
213. S. Romagnani et al, 1981, *J. Immunol.*, **127** 1307
214. S. Romagnani et al, 1981, *J. Immunol.*, **129** 596
215. T. Teranishi et al, 1984, *J. Immunol.*, **33** 3062
216. H. Harada et al, 1982, *Immunology*, **47** 557
217. A. Muraguchi et al, 1980, *J. Immunol.*, **125** 164
218. E. Smith et al, 1983, *J. Immunol.*, **130** 773

219. A. Ducatenzeiler and J. Menezes, 1988, *Int. J. Pharmacol.*, **10** 81
220. T. Hirano et al, 1984, *J. Immunol.*, **133** 798
221. A. Levinson et al, 1986, *J. Clin. Invest.*, **78** 612
222. T. Kasahara et al, 1981, *Immunology*, **42** 175
223. A. Sulica, 1976, *Scand. J. Immunol.*, **5** 1191
224. A. Sulica et al, 1979, *Immunology*, **38** 173
225. A. Sulica et al, 1979, *Eur. J. Immunol.*, **9** 979
226. A. Sulica et al, 1979, *Eur. J. Immunol.*, **9** 985
227. M. Laky et al, 1978, *Scand. J. Immunol.*, **7** 345
228. J. Dossett et al, 1969, *J. Immunol.*, **103** 1045
229. R. Harvey et al, 1970, *Proc. Soc. Exp. Biol. Med.*, **135** 453
230. J. Verhoef et al, 1977, *Immunology*, **33** 191
231. P. Peterson et al, 1972, *Infect. Immunol.*, **15** 760
232. S. Hsieh et al, 1978, *J. Infect. Dis.*, **138** 754
233. G. Gross et al, 1978, *Infect. Immun.*, **21** 7
234. G. Kronvall, 1973, *J. Immunol.*, **111** 1406
235. P. Pearlman et al, 1972, *J. Immunol.*, **108** 558
236. H. Lustig and C. Bianco, 1976, *J. Immunol.*, **116** 253
237. A. Levinson et al, 1987, *J. Immunol.*, **139** 2237
238. I. Lind and B. Mansa, 1974, *Acta Path. Microbiol. Scand., Sect. B.*, **82** 829
239. I. Lind and B. Mansa, *Scand. J. Immunol.*, **3** 147
240. F. Nardella and I. Oppliger, 1990, in Bacterial Immunoglobulin-Binding Proteins, (M. Boyle, ed.) Vol. I, p. 317, Academic Press, San Diego
241. D. Terman, 1981, *Fedn. Proc.*, **40** 45
242. P. Ray et al, 1979, *Fedn. Proc.*, **38** 1089
243. D. Kay et al, 1977, *J. Immunol.*, **118** 2058
244. R. Austin and C. Daniels, 1976, *J. Immunol.*, **117** 602
245. T. Jordan et al, 1978, *Cancer Res.*, **38** 3707
246. G. Shaw et al, 1978, *J. Immunol.*, **121** 573
247. Y. Kim et al, 1980, *J. Immunol.*, **125** 755
248. E. Ades et al, 1976, *J. Immunol.*, **117** 2119
249. M. Shibasaki et al, 1978, *J. Immunol.*, **121** 2278
250. J. Hawager et al, 1972, *J. Exp. Med.*, **136** 68
251. J. Hawager et al, 1979, *J. Clin. Invest.*, **64** 931
252. W. Bensinger et al, 1984, *J. Biol. Resp. Mod.*, **3** 347
253. G. Messerschmidt et al, 1984, *J. Biol. Resp. Mod.*, **3** 325
254. D. Terman and D. Bertram, 1985, *Eur. J. Cancer Clin. Oncol.*, **21** 1105
255. G. Ventura et al, 1987, *Cancer Treat. Rep.*, **71** 411
256. —1994, 2-Step Purification of a Mouse IgG_1 Monoclonal Antibody, Application Note 215, publication 18-1060-42, Pharmacia Biotech AB, Uppsala, Sweden
257. J. Sjöquist et al 1972, *Eur. J. Biochem.*, **29** 572
258. J. Movitz, 1976, *Eur. J. Biochem.*, **68** 291
259. R. Carlsson et al, 1984, *Cell. Immunol.*, **86** 136
260. P. Shaddie et al, 1995, Antibody Purification, US patent

5,429,746
261. C. Das et al, 1985, *Anal. Biochem.*, **145** 27
262. Intact Cowan strain and recombinant B domains are available from Sigma Chemical Company, St. Louis, MO,
263. H. McConnell et al, 1992, *Science*, **257** 1906
264. H. McConnell et al, 1995, *Proc. Nat. Acad. Sci.*, **92** 2750
265. M. Renschler et al, 1995, *Cancer. Res.*, **55** 5642
266. —1996, The CytoSensor Microphysiometer at Work, Molecular Devices Corporation, Sunnyvale, CA USA
267. A. Moschella, 1996, personal communication, Bioprocessing Inc., Princeton, NJ USA
268. J. Stephenson et al, 1974, *Anal. Biochem.*, **142** 189
269. Protein-A CellThru BIGBEADS, Sterogene Bioseparations Inc., Carlsbad, CA, USA
270. rProtein A STREAMLINE, Pharmacia Biotech AB, Uppsala, Sweden
271. PROSEP-A High Capacity, Bioprocessing Ltd., Consett, UK
272. M. Boyle (ed.), 1990, Bacterial Immunoglobulin-Binding Proteins, 2 Vols., Academic Press, San Diego

Chapter 10

Other Bioaffinity Chromatography Methods

"Curiouser and Curiouser."
—Lewis Carroll

With the exception of protein A, the use of biological affinity for purification of monoclonal antibodies has been limited. Protein G offers some minor advantages over protein A but it shares the majority of its weaknesses and has severe limitations of its own. Immunoaffinity can be conducted in a variety of formats that collectively satisfy a wide range of requirements. However, each format bears individual restrictions that limit its scope of application. Lectin affinity has been more of a research specialty than a serious process tool. Other biological ligands account for anecdotal usage only.

Protein G

Protein G is derived from group C and G streptococci.[1-4] It comprises a single polypeptide with multiple binding domains linked in a cylindrical conformation (Figure 10.1).[5,6] Wild-type protein G contains albumin-binding domains as well as IgG-Fc and Fab-binding domains.[7-11] The current generation of immobilized protein G products exclusively employ recombinant versions from which the albumin-binding

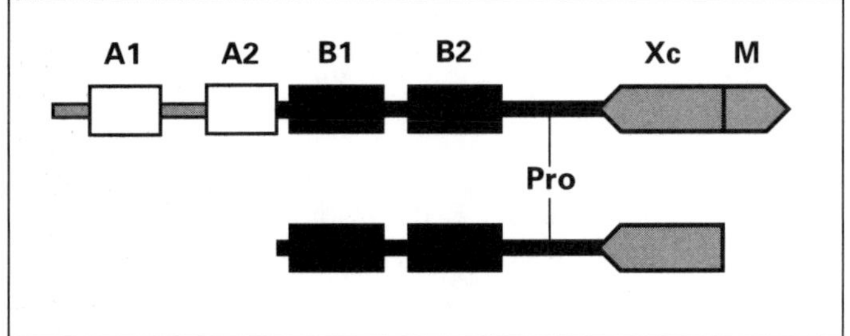

Figure 10.1. Domain structure of wild type (top) and recombinant protein G. The "A" domains bind albumin. The "B" domains bind the Fc region of IgG. "Pro" is a proline-rich sequence that binds IgG-Fab.Xc and M are the trans-cell-wall and membrane anchorage regions. See text for discussion.

domains have been deleted. Some recombinants lack the hydrophobic C-terminal sequence, the intent being to minimize nonspecific hydrophobic interactions.[12]

Protein G binds the same region of IgG-Fc as protein A, and they competitively inhibit one another.[10, 11,13] Despite this and their gross conformational similarity, there is no homology with respect to composition, number, arrangement, or functional characteristics of their IgG-binding domains.[14,15] Their amino acid sequences are distinct. Protein G Fc-binding affinity increases by a factor of 10 with each added domain.[7] Protein A Fc-binding is unaffected by the number of domains. Protein G's binding optimum is at acidic pH, versus protein A's alkaline optimum.[5,16-18] Protein G binds human IgG_3 of the Caucasian allotype G3m (s^-t^-) where protein A does not, and the two proteins are antigenically distinct.[18,19]

attributes

As with other bioaffinity methods, one of protein G's most attractive features is its simplicity. It requires no special chemicals, no special equipment, and only basic laboratory skills.

Protein G binds IgG more strongly than protein A. However, the monoclonals for which protein G shows the weakest affinity are also the ones for which protein A has the weakest affinity: mouse IgG_1s and rat IgGs.[20] Binding is strongest at pH 5.0 for all IgGs.

purification performance

Purity of total IgG is typically >90% but monoclonal purity may be as little as 50%, depending upon the content of nonspecific antibodies. There is no latitude to develop selective binding or elution conditions. Protein G binds α_2-macroglobulin and kinogen along with nonspecific IgG.[15]

mass and activity recovery

Mass recovery can be as high as 80–90% for mouse and rat monoclonals, or as low as 60% for human antibodies. This reflects the strong binding interaction. Harsh elution conditions often depress immunoreactivity of the harvested antibody, especially among human IgGs.[21]

DNA, virus, and endotoxin clearance

DNA, virus, and endotoxin clearance data have not been published, but it's reasonable to expect that they should be on par with protein A.

high capture efficiency

Protein G supports efficient capture of antibody from dilute feedstreams without buffer exchange or addition of binding enhancers, although capacity is improved by binding at pH 5.0.

limitations

Protein G's most serious limitation is its requirement for harsh elution conditions. Even mouse and rat IgG usually require pH 3.0, and human antibodies may require pH below 2.5. Immune function is seldom recovered without detectable alteration. Loss of titer is usually the most obvious symptom. Elevated nonspecific behavior, higher aggregate content, and depressed shelf stability are also frequent. These problems can be minimized with diligent care to neutralize fractions instantly upon elution, but not eliminated.

lot-to-lot product carryover

Strong IgG binding is disadvantageous in 2 additional respects. First, it maximizes the probability of lot-to-lot IgG carryover between runs. This mandates that separate columns be dedicated to individual antibodies. Second, uneluted IgG reduces product binding capacity for subsequent runs. It's common for protein G columns to lose 30-50% of their binding capacity after a few purifications. Unfortunately, protein G lacks the robustness of protein A versus harsh cleaning agents. The compound effect of these limitations is truncated column life.

leaching

As with protein A, and for the same reasons, leaching is a serious liability. The immunomodulatory effects of protein G are less characterized than those of protein A but they follow a similar pattern. This is to be expected, since they bind the same region of IgG. Besides lymphocyte proliferation, protein G has been reported to activate complement and mediate leukocyte chemotaxis.[22-26]

leachate removal

Leached protein G is far more difficult to remove than leached protein A because of its strong affinity. Insoluble protein G:IgG complexes can be filtered out and soluble complexes reduced by size exclusion (SEC) or ion exchange chromatography (IEC), but the extreme conditions required for complete dissociation make it impossible to recover native antibody quantitatively free of protein G.

variability among commercial products

Purification performance varies significantly among commercial protein G products.[14,22] Binding capacity may vary by a factor of 2–3. This partly reflects variations in ligand density, spacer characteristics and the like, but it also derives from variations in the protein G. Variations in IgG binding capability of protein G from different cell lines are sometimes paralleled by differences in their ability to stimulate proliferation of peripheral blood lymphocytes. Whether these variations reflect differences in the protein G itself or contaminant content has not been determined.

Protein G in perspective

Protein G offers some advantages over protein A. It is specific for IgG, it binds mouse and rat antibodies effectively without requiring individual method development, and it binds most antibodies from raw feedstreams without requiring alteration of pH or conductivity. However, the price for these conveniences is prohibitive. Lack of ability to remove nonspecific antibody, greater problems with denaturation, virtual impossibility of removing immunoreactive leachate, and higher compound operating expense are the key points that make it an inferior choice. Immobilized metal affinity chromatography (IMAC) offers broader applicability with fewer compromises than either.

Protein A/G hybrids

The one asset that recombinant protein A/G hybrids offer is that they bind a broader range of antibodies than either of the parent proteins.[27] They bind ruminant IgG strongly where protein A does not. They bind dog, cat, pig, and guinea pig IgG strongly where protein G does not. They offer nothing novel for process-scale purification of mouse, rat, and human monoclonals. Binding capacity is intermediate between the parent ligands. Elution characteristics are dominated by the protein G component. As with the individual ligands, IMAC is usually a superior alternative.

Other bacterial immunoglobulin-binding proteins

A number of other bacterial immunoglobulin-binding ligands have been described, although not commercialized (Table 10.1).[28-47] Protein B, derived from group B streptococci, is specific for the Fc region of human IgA.[28-30] Group B streptococci have also been reported to yield an IgG-binding protein (CAMP

Table 10.1. Specificities of bacterial immunoglobulin-binding proteins.

	IgA	IgD	IgE	IgG	IgM
Protein A	+	+	+	+++	++
Protein ARP	+++	–	–	+	–
Protein B	+++	–	–	–	–
Protein "D"	–	+++	–	+	+++
Protein G	–	–	–	+++	–
Protein L	++	++	++	++	+++
Protein P	++	–	–	++	+++

factor).[31] Protein ARP, derived from group A Streptococci, binds strongly to human IgA Fc, weakly to human IgG.[28,32-38]

An IgD-binding protein has been described from *Branhamella catarrhalis*.[39,40] It also binds small amounts of IgG and at least one report suggests that it binds IgM as effectively as as it binds IgD.[41] A protein with similar though weaker binding properties has been described from *Haemophilus influenzae*.[39] IgD binding proteins also occur in group A, C, and G streptococci, but they bind more IgG than IgD.[42]

Protein L is a light chain-binding protein derived from *Peptococcus magnus*.[43-45] Its gross conformational structure is similar to protein A and protein G. It binds all classes of human immunoglobulin but only those with kappa light chains. It binds IgM most strongly, apparently as a result of it having five times as many binding sites.[45] It binds rat and mouse antibodies but with lower affinity than human. Protein P, derived from *Clostridium perfringens*, displays selectivity similar to protein L.[46,47]

Immunoaffinity

Immunoaffinity is the prototype for biological affinity purification, and the most varied in terms of the formats in which it can be applied. There have been few reported applications, and most of those have been motivated by lack of convenient broad-spectrum affinity ligands for monoclonal antibody groupings regarded as "difficult," chiefly IgMs and rat IgGs.[48-53]

immobilized antigen affinity

The classical immunoaffinity format employs immobilized antigens. Despite its conceptual attractiveness, this approach has never developed wide appeal for monoclonal purification. The antigens are often not soluble, and among those that are, many occur at low concentrations and require arduous purification procedures. Most of these antigens are complex and may copurify antibodies with undesired specificities. Secondary steric modifications induced by immobilization may alter capture efficiency; even specificity.

Whatever procedures are required to prepare and immobilize the antigens are compounded by the need for periodic media replacement. Another concern with this approach is that binding and elution may alter a monoclonal's antigen-binding site. Leached antigen will interfere with the intended application. It's doubtful that any monoclonal product justifies the expense of overcoming all these difficulties.

immobilized monoclonal affinity

Similarly narrow specificity is provided by immobilizing a capture monoclonal that has been developed specifically for the target monoclonal. An additional advantage of this approach is that the target monoclonal can be selected to elute under moderate chemical conditions. The main limitation is that each target monoclonal requires development of a specific capture monoclonal. This approach is often used to purify high value recombinant proteins, but again, monoclonals generally don't justify the expense.

chain-specific affinity

Immobilized anti-light chain-specific antibodies copurify same-species nonspecific antibodies and free light chains indiscriminantly, but they offer a ready means for purification of non-IgG monoclonals.[54] Immobilized anti-heavy chain antibodies are widely available but not widely used. Narrow monoclonal specificities fail to recognize many target antibodies. Strong affinity of broadly-specific polyclonal antibodies requires excessively harsh elution conditions.

limitations

Immunoaffinity-based media are subject to the same validation requirements as the products they are used to purify. This compounds development costs for an already-compromised class of purification products.

Peptide and oligonucleotide affinity

Peptide and oligonucleotide-based affinity ligands promise a whole new spectrum of selectivities. These ligands have already begun to emerge commercially but they are difficult to identify; they are natural candidates for all manner of catchy but nondescriptive trade names. It remains to be seen how they will compete with established technologies.

Lectin affinity

Lectins are carbohydrate-binding proteins derived from a variety of biological sources. They have been known to bind to antibodies for over 25 years and have been available in immobilized form since the 1970s. However, their application to antibody purification has been sporadic. The technique has been neither systematically evaluated nor commercially developed for purification of monoclonals.

attributes

Lectin affinity chromatography can be conducted simply and effectively with the most basic or the most sophisticated equipment. It requires no special technique, and even though it employs a unique class of eluting agents, they are readily available from standard biochemical suppliers. The range of individual lectin affinities covers all antibody classes (Table 10.2).[55-84]

purification performance

Some lectins support excellent purification. Immobilized GS-1 yields >95% pure murine IgD.[60] Jacalin provides very good purity with human IgA, as does GNA with mouse IgM.[56,60-62,63-70] Lectins with specificities to widely distributed carbohydrates may capture more contaminants than antibody. Con A binds α_1-antiprotease, haptoglobin, transferrin, α_2-macroglobulin, β-lipoprotein, and nearly all the complement proteins, along with IgM and a minority of IgGs.[71-76] Besides limiting purity, this reduces the binding capacity available for the target monoclonal.

Antibody purification performance can be improved on broad-specificity lectins by employing gradient fractionation instead of single-step elutions. This has proven useful with immobilized RCA-1, where a multiple step gradient was used for separating human IgA_1 from IgA_2.[77]

DNA clearance

DNA, virus, and endotoxin clearance capabilities are uncharacterized. DNA clearance should be uniformly

Table 10.2. References reporting lectin affinity for antibodies.
Con A: concanavalin A
GNA: *Galanthus nivalis* agglutinin
GS-1: *Griffonia simplicifolia*
Jacalin: jackfruit lectin
LcH: Lentil lectin
MBP: mannan binding protein
PHA: *Phaseolus vulgaris* agglutinin
PNA: peanut agglutinin
PSA: *Pisum sativa* agglutinin
PWA: pokeweed agglutinin
RCA: ricin agglutinin
*Some of these references report interactions with glycopeptides and may not reflect the behavior of intact antibody. †Some report agglutination reactions in free solution and may not reflect the behavior of immobilized lectin.

	IgA	IgD	IgE	IgG	IgM
Con A	71	—	—	57,71-75	57,71-73,76
GNA	—	—	—	—	56
GS-1	—	60	—	—	—
Jacalin	60-62,63-70	60,63-67	61,62	—	61,62
LcH*	—	—	55	55,79,80	55,79
MBP	—	—	—	—	78
PHA†	81-83	83	—	81	81
PNA*	—	—	—	75	—
PSA*	—	—	—	55	—
PWA*	—	—	—	75	—
RCA-1	77	59	—	57,58,74,84	57,74,84
RCA-2	—	60	—	—	—

good. Viral and endotoxin clearance will likely prove to be individual lectin-dependent.

mass recovery

Mass recovery is typically >90% and activity recovery is usually quantitative, both owing to gentle separation conditions. The exception is PHA, which may require aggressive elution.

Limitations

Selective purification of post-translational glycosylation isoforms is a greater concern with lectin affinity than with any other purification technique. Variations in glycoform distribution affect antibody titer, stability, and pharmacokinetics.[75,85-90] IEF profiles of lectin-purified monoclonals should be cross-checked versus an unpurified control to ensure that the native proportions among glycosylation isoforms are maintained.

leaching

Ligand leakage is important because most lectins are potent immunomodulators. Effects include complement activation, stimulation of lymphocyte proliferation, induction of suppressor cells, toxicity of lymphocytes and macrophages, phagocytosis of target cells, insulinometric activity, and cellular toxicity.[91-93]

ligand heterogeneity

Lectin preparations are heterogeneous. Jacalin from the Philipines is reported to bind only IgA_1.[61] Indian and Brazilian jacalin bind IgA_1, IgA_2 and IgD.[63-80] Jacalin from Okinawa binds IgM and IgE as well.[61] It has not been determined whether these differences are attributable to genetic variation, local growth conditions, or both.

PWA represents another pattern of variation. It is composed of 5 isolectins, the proportions of which vary according to which part of the plant the lectin is harvested from and the plant's degree of maturity.[94] PSA and LcH exhibit a similar pattern.[94] Finally, not all lectin preparations are purified to an equal degree and, among those that are, activity may vary with the method of preparation.

limited availability

The majority of immobilized lectins are obtainable only on agarose, sometimes cross-linked, sometimes not. The chemistries employed for immobilization have not been selected in anticipation of manufacturing usage. Most lectins are multimeric and held together by noncovalent forces. This makes them labile to harsh sanitizing conditions and restricts their operating conditions in some cases as well.

cost

Immobilized lectins are expensive, mainly because they are manufactured and sold in small quantities. Their expense is compounded by their low capacity—usually 1–2mg/mL of gel—and short usage life.

Alkaline peptide ligands

Several alkaline polypeptides have been evaluated for IgM purification, including C1q, protamine, histone, and lysozyme.[95-97] Use of C1q was inspired by the efficiency of complement fixation by IgM.[95] Although it preferentially binds IgM, it also adsorbs IgG and has a very high affinity for aggregated IgG.[96] Its real purification capability remains unclear since the only reported applications followed ammonium sulfate precipitation. Immobilized protamine supports efficient IgM capture but purity is poor: ~40%.[97] The authors suggested that the interaction was localized to a dominantly anionic site on the antibody, but the overall selectivity was consistent with the protamine simply acting as an anion exchanger.

DNA affinity

IgM binds immobilized DNA in a perverse variant of cation exchange chromatography.[98] This should be regarded as a warning about free-solution interactions between monoclonals and contaminating DNA.

carbohydrate affinity

IgMs with specificities to carbohydrate antigens can sometimes be purified on immobilized sugars—a sort of reverse lectin affinity approach.[99] This is also a warning: some IgMs may bind immunospecifically to carbohydrate-based chromatography media.

Recommended reading

Michael Boyle's 2 volume series is an outstanding resource for information on bacterial immunoglobulin-binding proteins of all kinds.[100] Chapter 12, volume 2, specifically discusses monoclonal purification with protein G. A recent book by Hermanson is an excellent resource for people who wish to immobilize their own antibodies (or antigens).[27] References on antibody purification by lectin affinity are few, and in most cases purification is incidental to the subject. Begin with the references listed in Table 10.2.

References

1. K. Reis et al, 1984, *J. Immunol.*, **132** 3091
2. L. Björck and G. Kronvall, 1984, *J. Immunol.*, **133** 969
3. K. Reis et al, 1986, *Mol. Immunol.*, **23** 425
4. G. von Mehring and M. Boyle, 1986, *Mol. Immunol.*, **23** 811
5. B. Åkerström and L. Björck, 1986, *J. Biol. Chem.*, **261** 10240
6. A. Olsson et al, 1987, *Eur. J. Biochem.*, **168** 319
7. S. Fahnestock et al, 1990, in Bacterial Immunoglobulin-Binding Proteins, (M. Boyle, ed.), Vol. 1, p. 133, Academic Press, San Diego
8. J.-I. Jörnvall et al, 1986, *EMBO J.*, **5** 1567
9. M. Erntell et al, 1988, *Acta. Pathol. Microbiol. Scand. Sect. B*, **94** 377
10. M. Eliasson et al, 1989, *J. Immunol.*, **142** 575
11. A. Schröder et al, 1986, *Immunology*, **57** 305
12. J. McGuire, 1989, *Am. Biotech. Lab.*, **Sept.** 28
13. K. Reis et al, 1984, *J. Immunol.*, **132** 3098
14. K. Reis and M. Boyle, 1990, in Bacterial Immunoglobulin-Binding Proteins, (M. Boyle, ed.) Vol. 1, p. 101, Academic Press, San Diego
15. L. Björk and B. Akerstrom, 1990, in Bacterial Immunoglobulin-Binding Proteins, (M. Boyle, ed.) Vol. 1, p. 113, Academic Press, San Diego
16. J. Woof and D. Burton, 1990, in Bacterial Imunoglobulin-Binding Proteins, (M. Boyle, ed.), Vol. 1, p. 305, Academic Press, San Diego
17. B. Walker, 1990, in Bacterial Imunoglobulin-Binding Proteins, (M. Boyle, ed.), Vol. 2, p. 355, Academic Press, San Diego

18. G. Kronvall and R. Williams, 1969, *J. Immunol.*, **103** 828
19. M. Boyle and K. Reis, 1990, in Bacterial Immunoglobulin-Binding Proteins, (M. Boyle, ed.) Vol. 1, p. 175, Academic Press, San Diego
20. S. Ohlson and S. Gustafson, 1988, *BioMedia*, **1** 6
21. J. Tedesco et al, 1989, *BioChromatography*, **4** 216
22. E. Faulmann et al, 1989, *J. Immunol. Met.*, **123** 269
23. D. Barrett, 1990, in Bacterial Imunoglobulin-Binding Proteins, (M. Boyle, ed.), Vol. 1, p. 229, Academic Press, San Diego
24. M. Boyle, 1990, in Bacterial Imunoglobulin-Binding Proteins, (M. Boyle, ed.), Vol. 1, p. 295, Academic Press, San Diego
25. M. Boyle, 1990, in Bacterial Imunoglobulin-Binding Proteins, (M. Boyle, ed.), Vol. 2, p. 369, Academic Press, San Diego
26. M. Boyle et al, 1988 in Inflammatory Bowel Disease: Current Status and Future Approaches, (R. MacDermott,ed.), p. 371, Excerpta Medica, Amsterdam
27. G. Hermanson et al, 1992, Immobilized Afinity Ligand Techniques, Academic Press, San Diego
28. G. Russel-Jones et al, 1984, *J. Exp. Med.*, **160** 1467
29. L.-J. Brady and M. Boyle, 1990, in Bacterial Imunoglobulin-Binding Proteins, (M. Boyle, ed.), Vol. 1, p. 201, Academic Press, San Diego
30. P. Cleat and K. Timmis, 1990, in Bacterial Imunoglobulin-Binding Proteins, (M. Boyle, ed.), Vol. 1, p. 225, Academic Press, San Diego
31. D. Jürgens et al, 1987, *J. Exp. Med.*, **165** 720
32. G. Lindahl and B. Åckerström, 1989, *Molec. Microbiol.*, **3** 239
33. P. Christensen and V. Oxelius, 1975, *Acta Pathol. Microbiol. Scand. Sect. C.*, **83** 28
34. C. Schálen, 1980, *Acta Pathol. Microbiol. Scand. Sect. C.*, **88** 271
35. L. Burova et al, 1981, *Acta Pathol. Microbiol. Scand. Sect. C.*, **89** 433
36. G. Kronvall et al, 1979, *Infect. Immunol.*, **25** 1
37. C. Schálen et al, 1982, *Acta Pathol. Microbiol. Scand. Sect. B.*, **90** 347
38. G. Lindahl et al, 1990, in Bacterial Imunoglobulin-Binding Proteins, (M. Boyle, ed.), Vol. 1, p. 193, Academic Press, San Diego
39. T. Tedder,1990, in Bacterial Imunoglobulin-Binding Proteins, (M. Boyle, ed.), Vol. 1, p. 235, Academic Press, San Diego
40. A. Forsgren and A. Grubb, 1979, *J. Immunol.*, **122** 1468
41. A. Forsgren et al, 1980, *Scand. J. Infect. Dis.*, **24** (suppl.) 112
42. S. Vidal and S. Conde, 1980, *J. Immunol. Met.*, **36** 169
43. E. Myhre and M. Erntell, 1985, *Molec. Immunol.*, **22** 879

44. L. Björck and B. Åkerstöm, 1990, in Bacterial Immunoglobulin-Binding Proteins, (M. Boyle, ed.), Vol. 1, p. 267, Academic Press, San Diego
45. L. Björck, 1988, *J. Immunol.*, **140** 1194
46. G. Lindahl and G. Kronvall,1988, *J. Immunol.*, **140** 1223
47. G. Lindahl, 1990, in Bacterial Imunoglobulin-Binding Proteins, (M. Boyle, ed.), Vol. 1, p. 257, Academic Press, San Diego
48. P.-L. Lim, 1987, *Molec. Immunol.*, **24** 1
49. J. Groelke et al, 1987, *Hybridoma*, **6** 259
50. F. Cormont et al, 1986, *Met. Enzymol.*, **121** 622
51. Y. Meng and J. Trawinski, 1988, *J. Immunol.*, **141** 2684
52. H. Bazin et al, 1986, *Met. Enzymol.*, **121** 638
53. H. Bazin et al, 1984, *J. Immunol. Met.*, **71** 9
54. P. Grandics, 1994, *Am. Biotech. Lab.*, **Jun.** 12
55. K. Kornfield et al, 1981, *J. Biol. Chem.*, **256** 6633
56. N. Shibuya et al, 1988, *Arch. Biochem. Biophys.*, **267** 676
57. R. Klein et al, 1979, *Molec. Immunol.*, **16** 421
58. E. Saltvedt and J. Natvig, 1977, *Scand. J. Immunol.*, **6** 595
59. R. Marches and V. Ghetie, 1986, *Scand. J. Immunol.*, **24** 45
60. J. Oppenheim et al, 1990, *J. Immunol. Met.*, **130** 243
61. K. Hagiwara et al, 1988, *Molec. Immunol.*, **25** 69
62. H. Kondoh et al, 1986, *J. Immunol. Met.*, **88** 171
63. B. Zehr and S. Litwin, 1987, *Scand, J. Immunol.*, **26** 229
64. H. Kondoh et al, 1987, *Molec. Immunol.*, **24** 1219
65. P. Aucouturier et al, 1987, *Molec. Immunol.*, **24** 503
66. P. Aucouturier et al, 1989, *J. Clin. Lab. Anal.*, **3** 244
67. M. Roque-Barreira and A. Campos-Neto, 1984, *Bras. J. Med. Biol. Res.*, **17** 384
68. R. Gregory et al, 1987, *J. Immunol. Met.*, **99** 101
69. D. Skea et al, 1985, *Mol. Immunol.*, **25** 1
70. M. Roque-Barreira and A. Campos-Neto, 1985, *J. Immunol.*, **134** 1740
71. M. Leon, 1967, *Science*, **158** 1325
72. Y. Weinstein et al, 1972, *J. Immunol.*, **109** 1402
73. T. Bog-Hansen, 1973, *Anal. Biochem.*, **56** 480
74. E. Saltvedt et al, 1975, *Scand. J. Immunol.*, **4** 287
75. M. Malaise et al, 1987,*Clin. Immunol. Immunopathol.*, **45** 1
76. H. Harris and E. Robson, 1963, *Vox Sang.*, **8** 348
77. E. Saltvedt, 1976, *Scand. J. Immunol.*, **5** 1109
78. J. Nevens et al, 1992, *J. Chromatogr.*, **597** 247
79. M. Young and M. Leon, 1974, *Biochim. Biophys. Acta*, **365** 418
80. S. Kornfield et al, 1971, *J. Biol. Chem.*, **246** 6581
81. G. Spengler et al, 1972, *Schweiz. Med. Wschr.*, **102** 1618
82. H. Jaquet et al, 1975, *Z. Immune-Forsch.*, **150** 212
83. G. Spengler and R. Weber, 1979, *Protides Biol. Fluids*, **27** 615
84. E. Saltvedt et al, 1975, *Scand. J. Immunol.*, **4** (suppl. 2) 125

85. R. Parehk, 1992, *Biotech. Lab.*, **11** 61
86. Y. Kagawa, 1988, *J. Biol. Chem.*, **253** 508
87. B. Maiorella et al,1993, *Bio/Techniques*, **11**(3) 387
88. T. Monica et al, 1993, *Bio/Techniques*, **11**(4) 512
89. C. Goochee et al, 1990, *Bio/Techiques*, **8**(5) 421
90. C. Goochee et al, 1992, in Frontiers in Bioprocessing II, (P. Todd et al, eds.) p. 199, American Chemical Society, Washington, D.C.
91. H. Lis and N. Sharon, 1986, in The Lectins: Properties, Functions, and Applications in Biology and Medicine, (I. Liener et al, eds.), Academic Press, Orlando
92. J. Malin-Berdel et al, 1984, *Cytometry*, **5** 204
93. M. Ohta et al, 1990, *J. Biol. Chem.*, **265** 1980
94. —1992, Lectins and Lectin Conjugates, EY Laboratories, San Mateo, CA
95. A. Nethery et al, 1990, *J. Immunol. Met.*, **126** 57
96. J. Gibbons et al, 1981, *Biochim. Biophys. Acta*, **670** 146
97. A. Wichman and H. Borg, 1977, *Biochim. Biophys. Acta*, **490** 363
98. M. Abdullah et al, 1985, *J. Chromatogr.*, **347** 129
99. D. Marcus et al, 1989, *J. Immunol.*, **143** 2929
100. M. Boyle, M. (ed.), 1990, Bacterial Immunoglobulin-Binding Proteins, 2 Vols., Academic Press, San Diego

Appendix I

Foundation Protocols

"If a man will begin with certainties, he shall end in doubts; but if he will be content to begin with doubts, he shall end in certainties."
—Francis Bacon

The diversity of physicochemical characteristics among individual monoclonal antibodies is infinite, but in large part, that diversity represents variations on a theme. Monoclonals embody biochemical similarities that make it possible to apply standardized purification protocols with reasonably good success.

This appendix provides base protocols for a variety of applications. Many can be applied without modification for investigational purposes. Most require optimization for commercial applications. In vivo purifications require augmentation with supporting methods. Optimization strategies are detailed in the chapters within which the various methods are discussed.

Protocol 1. **Precipitation with octanoic acid followed by precipitation with ammonium sulfate.**

applicable to: most IgGs and IgYs, some IgAs and IgMs

purity, total Ab: >90%

purity, specific MAb: 70–90%

mass recovery, specific MAb: 50-80%

activity recovery per mass unit: 100%

most appropriate for: Small volumes. Investigational and in vitro product applications.

limitations: This protocol is optimized for purification of IgG from ascites and serum sources. Product recovery is generally low, especially with acidic antibodies and those with poor solubility at low pH and ionic strength. If residual octanoic acid is carried into the salt precipitation step it mediates secondary hydrophobic interactions that depress product solubility, elevate nonspecific interference, and reduce stability. This requires an intermediate complexant removal step. The overall process copurifies nonspecific IgG and traces of other immunoglobulins. Octanoic acid is malodorous.

materials: Buffer A: 0.05*M* sodium acetate, 0.05*M* sodium chloride, pH 4.0.

Buffer B: N-Octanoic acid

Buffer C: 1.0*M* Tris, pH 8.0

Buffer D: 0.1*M* Tris, pH 8.0

Buffer E: 3.0*M* ammonium sulfate, 0.1*M* sodium phosphate, pH 6.5

Buffer F: 0.05*M* sodium phosphate, 0.10*M* sodium chloride, pH 7.0

Dowex AG1X2, 400 mesh, chloride salt

method:
1. Bring all materials to room temperature (20–23°C).
2. Dilute sample to a standard protein concentration of ~10 mg/mL with buffer A. Adjust pH to 4.5.
3. Set sample stirring vigorously and add octanoic acid dropwise, 10µL per mL of sample. Stirring can be terminated after complexant addition. Incubate for 20–30 minutes.
4. Centrifuge at 10,000 x g for 15 minutes.
5. Decant then titrate the supernatant by addition of buffer C, 10% v:v. The change in pH will dissociate most of the residual octanoic acid from the antibody.
6. Equilibrate Dowex AG1X2 (400 mesh, chloride salt) to buffer D and add 1mL of settled gel per each 25 mL of supernatant. Incubate for at least 1 hour, stirring vigorously enough to keep the beads in suspension. The gel will scavenge free octanoic acid and maintain the steepest possible concentration gradient, favoring equilibrium dissocation of any IgG-associated precipitant.
7. Filter the supernatant through coarse filter paper. Discard the used gel media. It will be fouled irrecoverably.
8. Set the supernatant stirring and set up a peristaltic pump to deliver buffer E to a final concentration

of 1.85M ammonium sulfate so that the addition takes at least 15 minutes. Continue stirring for 30 minutes. If you prefer to do this step in the cold, cool all materials to temperature in advance then add buffer E to a final concentration of 2.0M ammonium sulfate.

9. Centrifuge at 10,000 x g for 20 minutes.
10. Resuspend precipitate with buffer F.

optimization: Purity and recovery can be improved by adjusting pH, conductivity, concentration of octanoic acid (or ammonium sulfate), and protein concentration of both steps (chapter 2). If antibody recovery is <70% after the octanoic acid step, try doubling the sodium chloride concentration of buffer A.

variations: Substituting ethacridine precipitation for the first step provides equivalent results (protocol 2). If you use ethacridine, then reduce the pH of the supernatant to <pH 7.5 before ammonium sulfate precipitation. Ammonia liberated at alkaline pH denatures the antibody. Substituting polyethylene glycol (PEG) precipitation for the second step gives equivalent results (protocol 4).

comments: This method and its variations are inexpensive, simple, and effective—excellent for small-volume purification without chromatography. The method described for ammonium sulfate addition dramatically reduces aggregate formation. It also works well for large-scale addition of PEG and other precipitants.

Protocol 2. Precipitation with ethacridine followed by size exclusion chromatography.

applicable to: Most IgGs and IgYs

purity, total Ab: >90%

purity, specific MAb: 70-90%

mass recovery, specific MAb: 60–80%

activity recovery per mass unit: 100%

most appropriate for: Small volumes. Investigational and in vitro product applications.

limitations: This protocol is optimized for purification of IgG from ascites and serum sources. Recovery is generally poor, especially with acidic antibodies. Ethacridine forms irreversible precipitates with inorganic anions, releasing lactic acid, driving down pH, and compromising purity. This requires advance removal of salts and sample reformulation with a strong organic buffer that can be titrated to pH without introduction of inorganic anions other than hydroxide. The overall process copurifies nonspecific IgG. Ethacridine is toxic.

materials: Buffer A: 0.05M BICINE, pH 8.0
Buffer B: 0.2M ethacridine in buffer A
Buffer C: 4.0M sodium chloride, 0.2M MES, pH 6.0
Buffer D: 0.05M sodium phosphate, 0.25M sodium chloride, 1.0M urea, pH 7.0
Size exclusion column of choice

method:
1. Bring all materials to room temperature.
2. Equilibrate sample to buffer A by method of choice.
3. Set sample stirring with a suspended stirrer or stir manually with a nonmetallic rod, and gradually add 1 part buffer B to 9 parts sample. Cease stiring after complexant addition. Incubate 20 minutes.
4. Decant the supernatant from the thick precipitate and filter it through coarse filter paper.
5. Add buffer C to a final concentration of 1.0M sodium chloride. This is to dissociate and precipitate residual ethacridine.
6. Equilibrate a size exclusion column with buffer D.
7. Filter the sample to 0.2µm and apply it to the column: ~5% v:v

optimization: Purity and recovery can be improved by adjusting pH, conductivity, protein concentration, and complexant concentration of the initial step (chapter 2). Urea can be omitted from the SEC buffer for many antibodies, but compare recovery and peak sharpness in advance (chapter 3). 1.0M urea does not constitute a denaturation risk.

variations: Improved fractionation of nonspecific antibodies can be achieved by replacing the size exclusion step with hydrophobic interaction chromatography (protocol 5).

comments: Ethacridine precipitation supports extraordinary clearance of DNA, virus, endotoxin, and lipids. It also yields extremely low aggregate populations.

Protocol 3. Precipitation with ammonium sulfate followed by anion exchange chromatography.

applicable to: all antibodies

purity, total Ab: >90%

purity, specific MAb: >90%

mass recovery, specific MAb: 50–70%

activity recovery per mass unit: 100%

most appropriate for: Small volumes. Investigational and in vitro product applications.

limitations: Low recovery from the precipitation step is a concern. This is a particular issue with cell culture supernatants because precipitation efficiency is depressed at low protein concentration. The anion exchange step requires linear gradient elution. Anion exchange exhibits poor capacity with strongly basic antibodies, especially weak exchangers like DEAE.

materials: Buffer A: 3.0M ammonium sulfate, 0.1M sodium phosphate, pH 6.5

Buffer B: 0.05M sodium phosphate, 0.1M sodium chloride, pH 7.0

Buffer C: 0.05M Tris, pH 8.5

Buffer D: 0.05M Tris, 1.0M sodium chloride, pH 8.5

Anion exchange media of choice

method:

1. Bring all materials to room temperature.
2. Conduct ammonium sulfate precipitation according to steps 8–10 from protocol 1.
3. Equilibrate sample to buffer C by method of choice.
4. Equilibrate anion exchange column with 10 column volumes (CV) buffer C.
5. Apply sample.
6. Wash with 2 CV buffer C.
7. Elute with a 10CV linear gradient to 25% buffer D.
8. Strip with 5 CV buffer D.

optimization: Both steps can be optimized to improve recovery and purity (chapters 2 and 4). The linear gradient can be converted to a step format once the intervals are defined. The process can be applied to non-IgG monoclonals and rat IgG, but purification is usually inferior and requires a third fractionation step.

variations: PEG precipitation can be substituted for the first step with equivalent results (protocol 4). It provides the advantage of lower salt concentration in the resuspended precipitate, which allows sample application to the ion exchanger without a buffer exchange step. Cation exchange can be substituted for anion exchange, often giving better purity, but anion exchange is more convenient because most antibodies remain fully soluble in the loading buffer.

comments: This protocol is the prototype for monoclonal IgG purification. Its remains useful after more than 20 years but it's not competitive with exclusively chromatographic methods.

Protocol 4. PEG precipitation followed by euglobulin adsorption/desorption.

applicable to: most IgMs

purity, total Ab: >90

purity, specific MAb: ~90%

mass recovery, specific MAb: 50–80%

activity recovery per mass unit: 100%

most appropriate for: Small volumes. Investigational and in vitro product applications.

limitations: Recovery is low, but not disproportionately so for IgM.

materials: Buffer A: 25% PEG-6000 in 0.05*M* sodium phosphate, 0.10*M* sodium chloride, pH 7.0
Buffer B: 0.05*M* sodium phosphate, 0.10*M* sodium chloride, pH 7.0
Buffer C: 0.01*M* HEPES, pH 7.0
Buffer D: 0.05*M* sodium phosphate, 1.5*M* sodium chloride, pH 7.0
Size exclusion media of choice, packed in a column with a bed height not greater than 15cm

method:

1. Bring all materials to room temperature.
2. Set the supernatant stirring and set up a peristaltic pump to deliver buffer A to a final concentration of 6% PEG, so that the addition requires ~15 minutes. Continue stirring for 30 minutes.
3. Centrifuge at 10,000 x g for 30 minutes.
4. Resuspend the precipitate with buffer B. Do not filter the sample

5. Equilibrate size exclusion media with 2CV buffer C. Flow rate: 20-50cm/hr.
6. Re-equilibrate the pump and lines up to the sample injector with Buffer D, but do not allow it to reach the column.
7. Apply sample (~5% CV) and continue with buffer D. The IgM will elute at the buffer D front.

optimization: Selectivity of PEG precipitation is more responsive to pH variation than salt precipitation. This makes it more optimizable but also means that results from generic conditions are variable from monoclonal to monoclonal. With some IgMs it may be possible to improve capacity by increasing column load to 20% of CV. See chapter 2 for low-salt variations of the euglobulin desorption buffer.

variations: The precipitation step can be applied to IgGs by increasing PEG concentration to 8%, but euglobulin adsorption works only with IgMs. Replacing PEG precipitation with hydrophilic interaction chromatography (HILIC) supports better recovery, purity, and scalability (chapter 8).

comments: Size-based methods are ordinarily unsuitable for removal of residual PEG, but in this case the SEC media is not being used to conduct SEC. It is acting as a nonspecific adsorbent, taking advantage of IgM's poor solubility. Residual PEG flows through with nonadsorbed contaminants.

Protocol 5. **Immobilized metal affinity followed by hydrophobic interaction chromatography.**

applicable to: most IgGs

purity, total Ab: >90%

purity, specific MAb: >90%

mass recovery, specific MAb: 80–90%

activity recovery per mass unit: 100%

most appropriate for: All scales. Investigational and in vitro applications, or as a platform for in vivo applications.

limitations: Requires a gradient chromatograph. IMAC copurifies nonspecific antibodies. Nickel is toxic and precipitates many antibodies. This requires affirmative metal ion control. Excessively hydrophobic supports reduce recovery and risk product denaturation. HIC buffers and conditions are designed for columns less hydrophobic than phenyl.

materials: Buffer A: 0.25M potassium phosphate, 2.5M sodium chloride, pH 8.5
Buffer B: 0.20M histidine, 0.05M EDTA, pH 7.0
Buffer C: 0.05M Tris, 0.05M EDTA, 0.50M sodium chloride, pH 8.0
Buffer D: 0.10M sodium acetate, 0.50M sodium chloride, pH 4.5
Buffer E: D plus 0.10M nickel chloride
Buffer F: 0.05M sodium phosphate, 0.5M sodium chloride, pH 8.0
Buffer G: 0.05M sodium phosphate 1.5M ammonium sulfate, pH 7.0
Buffer H: 0.05M sodium phosphate, 0.1M sodium chloride, pH 7.0
Immobilized imminodiacetic acid media of choice
HIC media of choice

method:
1. Bring all materials to room temperature.
2. Remove foulants from sample by method described in appendix II.
3. Equilibrate sample by adding 1 part buffer A to 4 parts sample.
4. Prespike fraction collection vessels with buffer B to 5% of their programmed volume.
5. Strip the iminodiacetic acid column with 2 CV buffer C.
6. Wash column with 2 CV buffer D.
7. Charge the column with nickel by loading buffer E until the bed is uniformly colored throughout, or monitor at 400nm until absorbance plateaus.
8. Wash off excess nickel with 5CV buffer D.
9. Equilibrate the column with 5CV buffer F.
10. Load ~10mg of antibody per mL of gel.
11. Wash with 2CV buffer F.
12. Elute in a 10CV linear gradient to buffer D.
13. Pool monoclonal-positive fractions.
14. Equilibrate HIC column with 80% buffer G, 20% buffer H.
15. Load ~10mg antibody per mL of gel by on-line

dilution, 20% sample, 80% buffer G.

16. Wash with 2CV, 80% buffer G, 20% buffer H.
17. Elute in a 10CV linear gradient to 100% buffer H.
18. Strip with 5CV 100% buffer H.
19. Pool monoclonal fractions.

optimization: Reducing IMAC binding pH will diminish contaminant binding, but also reduce IgG binding capacity. Excessive IgG losses during IMAC loading may indicate chelating agents in the column feed. Reverse the process order, using potassium phosphate for the HIC binding buffer.

variations: The IMAC sample equilibration step is optional but improves purity and recovery by disrupting ionic complexes between IgG and acidic contaminants. It also improves binding specificity, thereby increasing product capacity. For phenyl columns, reduce the salt concentration of buffer G to 1.0M. The cation exchange step from protocol 6—minus the protein A-dissociating wash—can be substituted for HIC. In this case, reduce the sodium chloride content of the IMAC post-load wash and elution buffers to 0.10M. Product may elute at a higher pH.

comments: IMAC provides most of the advantages of affinity capture without the disadvantages of biological ligands. The fraction spiking formulation is designed to scavenge and remove nickel that may have leached from the IMAC column. Purification with IMAC alone may be sufficient for antibodies produced in cell culture. IMAC can also be used for removal of contaminating IgG from non-IgG antibodies.

Protocol 6 Protein A affinity chromatography followed by cation exchange chromatography

applicable to: most IgGs, some IgMs and IgAs; not rat IgG_{2a}

purity, total Ab: >90%

purity, specific MAb: >90%

mass recovery, specific MAb: 70–80%

activity recovery per mass unit: 80-100%

most appropriate for: All scales. Investigational and in vitro applications, or as a platform for in vivo applications.

limitations: This protocol is designed for purification of Human and Mouse IgG. It requires a gradient chromatograph. The cation exchange step is critical where leached protein A may interfere with the product application. The cation exchange will require individual optimization to support effective protein A removal. Activity recovery may be reduced with Human IgG due to the harsh conditions required for dissociation from protein A.

materials:

- Buffer A: 0.02M citric acid, 0.50M sodium chloride, pH 3.0
- Buffer B: 1.0M TRIS, pH 7.5
- Buffer C1: 0.02M sodium phosphate, 0.02M sodium citrate, 0.50M sodium chloride, pH 7.5
- Buffer C2: 0.02M Boric acid, 0.02M sodium phosphate, 0.02M sodium citrate, 1.0M sodium sulfate, pH 9.0
- Buffer D: 0.02M Tris, 0.05M sodium chloride, pH 7.0
- Buffer E: 0.05M acetic acid, adjusted to pH 4.5 with sodium hydroxide
- Buffer F: 1.0M acetic acid, adjusted to pH 4.0 with sodium hydroxide
- Buffer G1: 0.05M acetic acid, 1.0M urea, 25% ethylene glycol, pH 4.5
- Buffer G2: 0.05M citric acid, 2.0M urea, 50% ethylene glycol, pH 3.0
- Buffer H: 0.05M acetic acid, 1.0M sodium chloride, pH 4.5
- Protein A media of choice
- Cation exchange media of choice

method:

1. Bring all materials to room temperature
2. Remove foulants from sample as described in appendix II
3. Pre-strip protein A column with 5CV buffer A
4. Prefill protein A collection vessels to 10% of programmed fraction volume with buffer B.

5a. For Human IgG: Equilibrate protein A column with 10CV buffer C1.

5b. For Mouse IgG: Equilibrate protein A column with 10CV buffer C2.

6a. For Human IgG: load up to 15mg IgG per mL of

gel. Pre-equilibration to buffer C1 is optional.

6b. For nonhuman antibodies: Use on-line dilution to equilibrate and load up to 5mg of antibody per mL of gel: 80% buffer B2, 20% sample.
7. Wash with 2CV buffer C.
8. Elute in a 10CV linear gradient from 100% C to 100% A. Pool monoclonal-positive fractions. Strip column with 5CV buffer A.
9. Prefill cation exchange collection vessels to 5% of programmed fraction volume with buffer B.
10. Equilibrate cation exchanger with 10CV buffer E.
11. Equilibrate harvested antibody to buffer D by method of choice.
12. Immediately before loading, pre-titrate the pH of the monoclonal pool by adding 5% (v:v) buffer F.
13. Use on-line dilution to equilibrate and load up 10mg antibody per mL of gel: 20% sample, 80% buffer E.
14. Wash with 1CV buffer E.
15. Wash with 10CV buffer G1 for Mouse antibodies, G2 for Human antibodies. This step is to dissociate product-bound protein A.
16. Wash with 2CV buffer E.
17. Elute in a 10 CV linear gradient to 25% buffer H
18. Strip with 5 CV 100% buffer H.

optimization: Every aspect of this protocol can be optimized. Gradient screening is necessary for both the protein A and cation exchange steps to identify appropriate elution conditions. Both can be subsequently converted to step gradients. Refer to chapter 9 for detailed discussion of leached protein A removal.

variations: Anion exchange supports fractionation of nonspecific antibodies similar to cation exchange, but is less effective for removing leached protein A.

comments: The 0.5 M sodium chloride in Human IgG binding buffer is optional. It is included to dissociate electrostatic complexes of antibodies with contaminants. The high pH and 1.0M sodium sulfate in the nonhuman antibody binding buffer are to enhance weak affinity with protein A and to dissociate electrostatic complexes. The long column equilibration step is to ensure complete pH titration of critical protein A histidyl residues (chapter 9). The phosphate/citrate buffer system is designed to give reproducible linear pH gradients. The 0.5M sodium chloride in the elution buffer is to enhance antibody solubility at low pH. It can be reduced, but not below 0.1M. Pretitration of fraction collection vessels is to prevent product denaturation.

Protocol 7. Ion exchange with hydrophobic interaction chromatography.

applicable to: all antibodies

purity, total Ab: >90

purity, specific MAb: >95%

mass recovery, specific MAb: 70–80%

activity recovery per mass unit: 100%

most appropriate for: All scales. Investigational and in vitro applications, or as a platform for in vivo applications.

limitations: This protocol requires a gradient chromatograph. Excessively hydrophobic supports reduce recovery and risk product denaturation.

materials and methods: HIC per protocol 5.
Anion exchange per protocol 3.
Cation exchange per protocol 6, minus the protein A dissociation step.

variations: Substituting hydroxyapatite chromatography (HAC) for either IEC method provides equivalent performance (chapter 5). Substituting thiophilic adsorption for HIC yields equivalent purification but elevated risk of product denaturation (chapter 8).

comments: Cation exchange followed by HIC is one of the most effective combinations of IEC/HIC. Most contaminants pass through the cation exchanger, leaving maximum capacity for the product. HIC can follow directly without buffer exchange.

HAC followed by HIC can also be very efficient. Many antibodies can be loaded on hydroxyapatite columns without on-line dilution—usually required for cation exchange. The relative weakness of HAC is that it binds a higher contaminant load. This compromises product binding capacity, which requires using a larger bed volume.

Appendix II

Sample Preparation

"There are no shortcuts to anyplace worth going."
—Beverly Sills

Column fouling by crude feedstreams is a serious detractor from both short and long-term process economies. It affects adsorbent capacity and purification performance even within a single run. It also reduces column life, no matter how rigorous your cleaning procedures. Advance defouling is essential.

filtration

Most chromatographers at least filter raw samples through 0.2µm membranes. This prevents fouling by gross cell debris. Some pass raw sample through depth filters of buffer exchange chromatography media. This has been reported to increase first-step column life by a factor of 10.[1] A growing population exploits one of the greatest gifts in the field of antibody purification.

adsorptive foulant removal

The "gift" is the charge differential between the majority of monoclonal antibodies and the worst column foulants. Most antibodies are strongly electropositive. Most foulants are either negatively charged or associated with foulants that are negatively charged (Table II.1).[2-8] Even though cholesterol and triglycerides are nonionic, they are almost exclusively associated with lipoproteins. Lipoproteins are highly electronegative by virtue of their phospholipid coats.

This provides a window of opportunity for selective foulant removal by anion exchange. Table II.2 illustrates foulant removal on several anion exchangers at physiological pH and ionic strength. IgAs and IgMs bind to varying degrees under these conditions. Most IgGs don't, or do so only weakly. For those that do, binding can be suspended by adding modest amounts of sodium chloride.

Table II.1. Major chromatography media foulants.

Foulant	character
Cell debris	negative, hydrophobic
DNA	negative
endotoxin	negative, hydrophobic
lipoproteins	negative, hydrophobic
phospholipids	negative, hydrophobic
fatty acids	negative, hydrophobic
triglycerides	hydrophobic
cholesterol	hydrophobic
phenol red	negative, hydrophobic

defouling formats

Defouling can be conducted in several formats, each suited to a different set of processing requirements. The simplest is to add microgranular cellulosic anion exchange media directly to the raw production medium. The exchanger should be equilibrated in advance to 0.05M sodium phosphate, 0.1M sodium chloride, pH ~7.2, or a rough equivalent. The amount of ion exchange media varies according to feedstream composition. Begin with ~5mL of moist cake per 100mL of sample. Incubate stirring for at least 1 hour. Overnight incubation in the cold is more effective. Filter the supernatant as usual. Expect filtration efficiency to improve by a factor of 5–10. Discard the fouled media. It won't be worth the trouble of cleaning.

A slightly more demanding approach is to prepare a depth filter by packing a short wide (or radial flow) column with a high-flow macroporous anion exchanger. This can be used as an in-line prefilter, or off-line. If it's off-line you can put a 0.22μm membrane filter downstream from the anion exchanger. The risk with putting the assembly on a chromatogaph is that in most cases it will be upstream of the pump. The pressure drop may cause solvent outgassing.

Neither of these formats are attractive for use with fluidized bed or Big-Bead technologies. Their advantage is their ability to accommodate debris-loaded samples. Centrifugal or filtrative pretreatment steps defeat their whole purpose. On the other hand, these media

Table II.2. Foulant clearance by various anion exchangers. Matrix 1 and 2 are macroporous particulate ion exchangers. Matrix 3 and 4 are microporous particulate exchangers. Matrix 5 is a microcrystalline nonporous medium. Cell culture supernatant was used as a control. LP: lipoprotein. CH: cholesterol. TG: triglyceride. PL: phospholipid. EU: endotoxin. PR: phenol red. All values in mg/mL except endotoxin (EU/mL) and DNA (pg/mL).

Sample	LP	CH	TG	PL	DNA	EU	PR
control	1.74	0.27	0.58	0.61	6×10^{8}	233	0.01
matrix 1	0.00	0.00	0.17	0.02	1×10^{5}	17	0.00
matrix 2	0.00	0.00	0.19	0.00	9×10^{4}	12	0.00
matrix 3	1.51	0.17	0.54	0.56	3×10^{8}	187	0.00
matrix 4	1.43	0.18	0.39	0.39	3×10^{8}	137	0.00
matrix 5	0.00	0.00	0.21	0.00	5×10^{4}	4	0.00

are more vulnerable to fouling than any other because they are exposed to cruder feedstreams. Fouling can and does cause problems even within the course of a single run. The solution is to use a prefilter of anion exchange CellThru BigBeads.[9] Particulate contaminants still pass through but soluble foulants are removed in advance of the main column.

The inlet frit of the anion exchange Big-Bead precolumn will foul with the same frequency that frits of crude-capture columns normally foul. However, this highlights another advantage. When the inlet frits of crude-capture columns foul, it normally causes flow disturbances that disrupt product capture efficiency. By putting a precolumn on an isolatable inlet line you protect the capture column from these disturbances. In most cases you'll be able to clear the obstruction simply by backflushing at an elevated flow rate.

Whichever format you pursue, its important to qualify performance in advance with small scale pilot experiments. Begin with raw sample plus ion exchange media and assay for antibody loss from the the treated sample. If recovery is less than 95%, add sodium chloride to a final concentration of ~0.05M, then re-evaluate recovery. If recovery is elevated to ~95%, then run another experiment with half the amount of sodium chloride. If recovery is still low, double the salt. Refine the range as necessary keeping in mind that the less salt, the more effective the process.

No matter what format you use, anion exchange-based defouling provides dividends beyond protecting your separation media. "Prepurification" removal of DNA and endotoxins can only improve the efficiency of the rest of your process for reducing these contaminants to target levels. It's likely that this treatment supports significant virus removal as well

References

1. R. Broeze et al, 1996, Development of Production and Purification Processes for a Monoclonal-Based Imaging Product, slide presentation, Waterside Monoclonal Conference, Norfolk
2. C. Prior, 1990, in Animal Cell Bioreactors, (C. Ho and D. Ng, eds.), p. 445, Butterworth, New York
3. M. Butler and M. Dawson (eds.) 1992, Cell Culture LabFax, Academic Press
4. H. Schultze and J. Heremans, 1966, The Molecular Biology of Human Proteins, Vol. 1, Elsevier, New York
5. —1995, Ex-Cyte Growth Enhancement Media Supplement, Bayer Corporation, Kankakee, IL USA
6. —1990, Art to Science in Tissue Culture, 9(4) 3, Hyclone Laboratories, Logan, UT USA
7. G. Sofer and L.-E. Nyström, 1991, Process Chromatography, A Guide to Validation, Academic Press, New York
8. M. Weary and F. Pearson, 1988, BioPharm, 1(4) 22
9. Q-CellThru BigBeads, Sterogene Bioseparations, Carlsbad, CA USA

Index

Abbreviations

Å	Angstrom
A_{280}	UV absorbance at 280nm
abs	Spectrophotometric absorbance
ACS	American Chemical Society
ALA	Alanine
ARG	Arginine
ASN	Asparagine
ASP	Aspartic acid
AU	Spectrophotometric absorbance units
Ba^{2+}	Barium
BSA	Bovine serum albumin
BICINE	Bis(hydroxyethyl)glycine
Br^-	Bromide
Ca^{2+}	Calcium
C1	Complement protein C1
C1q	Complement protein C1q
C3	Complement protein C3
C4	Complement protein C4
Cl^-	Chloride
ClO_4^-	Perchlorate
CM	Carboxymethyl
Con A	Concanavalin A
COO^-	Acetate
Cs^+	Cesium
CTG	Chymotrypsinogen
CV	Column volume
CYS	Cysteine
DEAE	Diethylaminoethyl
EDTA	Ethylenediamine tetraacetic acid
E-MULV	Ecotropic Murine leukemia virus
GLU	Glutamic acid
GLN	Glutamine
GLY	Glycine
GNA	*Galanthus nivalis* agglutinin
GS-1	*Griffonia simplicifolia* agglutinin
HAC	Hydroxyapatite chromatography
HBV	Hepatitus B virus

HDL	High density lipoprotein
HEPES	Hydroxyethylpiperazine ethanesulfonic acid
HETP	Height equivalent of a theoretical plate
HIC	Hydrophobic interaction chromatography
HILIC	Hydrophilic interaction chromatography
HIS	Histidine
HSV	*Herpes simplex* virus
IEC	Ion exchange chromatography
IEF	Isoelectric focusing
IDA	Iminodiacetic acid
ILE	Isoleucine
IMAC	Immobilized metal affinity chromatography
K^+	Potassium
LcH	Lentil lectin
LDL	Low density lipoprotein
LEU	Leucine
Li^+	Lithium
LYS	Lysine or Lysozyme
MBP	Mannan binding protein
MES	Morpholinoethanesulfonic acid
MET	Methionine
mEq	Milli-equivalents
Mg^{2+}	Magnesium
M_r	Molecular weight
MW	Molecular weight
Na^+	Sodium
NH_4^+	Ammonium
Ni-IDA	Nickel-charged iminodiacetic acid
NO_3^-	Nitrate
PAGE	Polyacrylamide gel electrophoresis
PBS	Phosphate buffered saline, usually a formulation approximating 0.05M sodium phosphate, 0.1M sodium chloride, pH ~7.0
PEG	Polyethylene glycol
PHA	*Phaseolus vulgaris* agglutinin
PHE	Phenyl or phehylalanine
pI	Isoelectric point
PNA	Peanut agglutinin
PO_4^{3-}	Phosphate
PPM	Parts per million
PRO	Proline

PSA	*Pisum sativa* agglutinin
PWA	Pokeweed agglutinin
QAE	Quarternary aminoethyl
Rb^+	Rubidium
RCA	Ricin agglutinin
RPE	R-phycoerythrin
R_s	Resolution
SCN^-	thiocyanate
SDS	Sodium dodecy sulfate
SEC	Size exclusion chromatography
SER	Serine
SO_4^-	Sulfate
SOP	Standard operating procedure
SP	Sulfopropyl
THR	Threonine
TRP	Tryptophan
Tris	Trishydroxyaminomethane
TYR	Tyrosine
USP	US Pharmacopeia
VAL	Valine
V_e	Elution volume
V_m	Matrix volume
V_o	Void volume
V_p	Pore volume
V_t	Column volume, total volume
X-MULV	Xenotropic Murine leukemia virus

Endnotes

The author and publisher invite your comments about this book. Plese send them by e-mail to: <ptmabs@validated.com>. Feel free to note errors, alternative points of view, and observations you think might benefit a second edition. Information about new literature references and techniques is always welcome and gratefully acknowledged.

About the author

"...Either write things worthy reading, or do things worth the writing."
—Benjamin Franklin

Pete Gagnon is President and Scientific Director of Validated Biosystems Inc., a bioprocess engineering firm specializing in downstream process development. During the past 15 years he has assisted over 3 dozen biotechnology companies worldwide—ranging from start-ups to multinational pharmaceutical corporations—to bring over 250 diagnostic and therapeutic products to market. The majority of these products have been based on monoclonal antibodies.

Pete also works with manufacturers of purification products to develop new purification technologies. He has been involved in all stages of product evolution with major chromatography media and systems suppliers worldwide. In addition to hands-on evaluation of emerging products, he has developed applications literature to help users derive the greatest benefits from these new resources.

The balance of Pete's time is dedicated to training and education. He has conducted technology training tours in the United States and Europe, made numerous presentations at bioprocess symposia, and published a range of materials on practical biomanufacturing issues. These publications include articles in journals such as J. Chromatography, LC-GC, and BioPharm. He is the Editor of Validated Biosystems Quarterly Resource Guide for Downstream Processing, a free Internet Newsletter. In addition to articles on various downstream processing themes, this website contains downloadable reprints of most of his publications. <http://www.validated.com>

As with downstream processing, he has an endless fascination for quotations, more of which will appear in his upcoming book, "Managing Upstream Contaminants in Downstream Processing."